Communications in Computer and Information Science   **614**

Commenced Publication in 2007
Founding and Former Series Editors:
Alfredo Cuzzocrea, Dominik Ślęzak, and Xiaokang Yang

Editorial Board

More information about this series at http://www.springer.com/series/7899

Tristan Cazenave · Mark H.M. Winands
Stefan Edelkamp · Stephan Schiffel
Michael Thielscher · Julian Togelius (Eds.)

Computer Games

Fourth Workshop on Computer Games, CGW 2015
and the Fourth Workshop on General Intelligence
in Game-Playing Agents, GIGA 2015
Held in Conjunction with the 24th International Conference
on Artificial Intelligence, IJCAI 2015
Buenos Aires, Argentina, July 26–27, 2015
Revised Selected Papers

 Springer

Editors

Tristan Cazenave
Université Paris-Dauphine
Paris
France

Mark H.M. Winands
Maastricht University
Maastricht
The Netherlands

Stefan Edelkamp
Universität Bremen
Bremen
Germany

Stephan Schiffel
Reykjavik University
Reykjavik
Iceland

Michael Thielscher
The University of New South Wales
Sydney, NSW
Australia

Julian Togelius
New York University
Brooklyn, NY
USA

ISSN 1865-0929   ISSN 1865-0937   (electronic)
Communications in Computer and Information Science
ISBN 978-3-319-39401-5   ISBN 978-3-319-39402-2   (eBook)
DOI 10.1007/978-3-319-39402-2

Library of Congress Control Number: 2016939911

Printed on acid-free paper

This Springer imprint is published by Springer Nature
The registered company is Springer International Publishing AG Switzerland

Preface

These joint proceedings contain the papers of the Computer Games Workshop (CGW 2015) and the General Intelligence in Game-Playing Agents (GIGA 2015) workshop, which were both held in Buenos Aires, Argentina. These workshops took place on July 26 and 27, 2015, respectively, in conjunction with the 24th International Conference on Artificial Intelligence (IJCAI 2015). These two workshops reflect the large interest in AI research for games.

The Computer and Games Workshop series is an international forum for researchers interested in all aspects of artificial intelligence (AI) and computer game playing. Earlier workshops took place in Montpellier, France (2012), Beijing, China (2013), and Prague, Czech Republic (2014). For the fourth edition of the Computer Games Workshop, 16 submissions were received in 2015. Each paper was sent to two reviewers. In the end, 10 papers were accepted for presentation at the workshop, of which eight made it into these proceedings. The published papers cover a wide range of topics related to computer games. They collectively discuss eight abstract games: Chinese checkers, Go Fish, Lost Cities, Morpion Solitaire, Phantom Domineering, Phantom Go, Settlers of Catan, and Surakarta. Additionally, one paper is on a roguelike game and one paper is on the Pancake Problem.

The GIGA workshop series has been established to become the major forum for discussing, presenting, and promoting research on general game playing (GGP). It aims at building intelligent software agents that can, given the rules of any game, automatically learn a strategy for playing that game at an expert level without any human intervention. The workshop intends to bring together researchers from subfields of AI to discuss how best to address the challenges and further advance the state of the art of general game-playing systems and generic artificial intelligence. Following the inaugural GIGA Workshop at IJCAI 2009 in Pasadena (USA), follow-up events took place at IJCAI 2011 in Barcelona (Spain) and IJCAI 2013 in Beijing (China). This fourth workshop on General Intelligence in Game-Playing Agents received 11 submissions. Each paper was sent to two reviewers. In the end, 10 papers were accepted for presentation at the workshop, of which four made it into these proceedings. The accepted papers cover topics such as general video game playing, advanced simulation-based methods, heuristics, and learning.

In all, 44 % of the submitted papers for both workshops were selected for these proceedings. Here we provide a brief outline of the 12 contributions, in the order in which they appear in the proceedings. They are divided into two parts: the first eight belong to the Computer Games Workshop and the last four to the GIGA Workshop.

Computer Games Workshop

"Challenges and Progress on Using Large Lossy Endgame Databases in Chinese checkers," written by Nathan Sturtevant, discusses using large endgame databases to improve the performance of minimax and Monte Carlo tree search (MCTS)-based agents in Chinese checkers. Several challenges are faced in how to properly integrate the endgame databases and how to correct errors that occur because of the compression that is used when storing the endgame data. Experimental results suggest that minimax-based approaches are able to do a better job of using the endgame data than MCTS approaches.

"Sequential Halving for Partially Observable Games," authored by Tom Pepels, Tristan Cazenave, and Mark Winands, investigates sequential halving as a selection policy in the following four partially observable games: Go Fish, Lost Cities, Phantom Domineering, and Phantom Go. Additionally, H-MCTS is studied, which uses sequential halving at the root of the search tree, and UCB elsewhere. Experimental results reveal that H-MCTS performs the best in Go Fish, whereas its performance is on par in Lost Cities and Phantom Domineering. Sequential halving as a flat Monte Carlo search appears to be the stronger technique in Phantom Go.

"An Experimental Investigation on the Pancake Problem," by Bruno Bouzy, discusses the pancake problem. It is an NP-hard problem and linked to the genome rearrangement problem also called sorting by reversals (SBR). To date, the best theoretical R-approximation was 2 with an algorithm, which gives a 1.22 experimental R-approximation on stacks with a size smaller than 70. In this paper a Monte Carlo search (MCS) approach with nested levels and specific domain-dependent simulations is used. The paper shows that MCS is an alternative to iterative deepening depth first search for sorting large stacks of pancakes. At a given level and with a given number of polynomial-time domain-dependent simulations, MCS is a polynomial-time algorithm as well. MCS at level 3 gives a 1.04 experimental R-approximation, which is a breakthrough. At level 1, MCS solves stacks of size 512 with an experimental R-approximation value of 1.20.

"485 – A New Upper Bound for Morpion Solitaire," a joint collaboration by Henryk Michalewski, Andrzej Nagórko, and Jakub Pawlewicz, shows a new upper bound of 485 moves for the 5T variant of the Morpion Solitaire game. This is achieved by encoding Morpion 5T rules as a linear program and solving 126,912 instances of this program on special octagonal boards. To show the correctness of this method, the rules of the game have been analyzed and the potential of a given position has been used. By solving continuous-valued relaxations of linear programs on these boards, an upper bound of 586 moves is obtained. Further analysis of original, not relaxed, mixed-integer programs leads to an improvement of this bound to 485 moves. However, this is achieved at a significantly higher computational cost.

"Multi-Agent Retrograde Analysis," by Tristan Cazenave, proposes a new predator–prey game. This domain is modeled as a board game where three predators are trying to capture a prey. Each agent has five possible moves: going up, down, left, right, or staying in the same location. The game terminates if the prey is on the same location as a predator or if the prey cannot move to an empty location. Small boards up to 9×9 have been solved using retrograde analysis. The outcome is that the predator–prey game is always lost for the prey when there are at least three predators.

"The Surakarta Bot Revealed," by Mark Winands, presents the ideas behind the agent SIA, which won the Surakarta tournament at the ICGA Computer Olympiad five times. The paper first describes SIA's αβ-based variable-depth search mechanism. Enhancements such as multi-cut forward pruning and realization probability search improve the agent considerably. Next, features of the static evaluation function are discussed as well. Experimental results indicate that features, which reward distribution of the pieces and penalize pieces that clutter together, give a genuine improvement in the playing strength.

"Learning to Trade in Strategic Board Games," written by Heriberto Cuayáhuitl, Simon Keizer, and Oliver Lemon, describes a data-driven approach for automatic trading in the game of Settlers of Catan. Their experiments are based on data collected from human players trading in text-based natural language. The performance of Bayesian networks, conditional random fields, and random forests have been compared in the task of ranking trading offers, and are evaluated both in an offline setting and online while playing the game against a rule-based baseline. Experimental results show that agents trained from data from average human players can outperform rule-based trading behavior, and that the random forest model achieves the best results.

"Argumentative AI Director Using Defeasible Logic Programming," a joint effort by Ramiro Agis, Andrea Cohen, and Diego Martínez, presents a novel implementation of an AI director that uses argumentation techniques to decide dynamic adaptations in the level generation of a roguelike game called *HermitArg*. The architecture of the game introduces *smart items* with defeasible information to be analyzed in a dialectical process.

GIGA Workshop

"On the Cross-Domain Reusability of Neural Modules for General Video Game Playing," written by Alex Braylan, Mark Hollenbeck, Elliot Meyerson, and Risto Miikkulainen, considers a general approach to knowledge transfer in which an agent learning with a neural network adapts how it reuses existing networks as it learns in a new domain. This approach is domain-agnostic and requires no prior assumptions about the nature of task relatedness or mappings. The method's performance and applicability are analyzed in high-dimensional Atari 2600 general video game playing.

"The GRL System: Learning Board Game Rules with Piece-Move Interactions," written by Peter Gregory, Henrique Coli Schumann, Yngvi Björnsson, and Stephan Schiffel, studies the problem of learning formal models of the rules of board games, using as input only example sequences of the moves made in playing those games. This work is distinguished from previous work in this area in that the interactions are learned between the pieces in the games. A previous game rule acquisition system is supplemented by allowing pieces to be added and removed from the board during play, and using a planning domain model acquisition system to encode the relationships between the pieces that interact during a move.

"Creating Action Heuristics for General Game Playing Agents," authored by Michal Trutman and Stephan Schiffel, investigates an approach that learns online heuristics that guide the simulations of MCTS in GGP. This approach generates heuristics that estimate the usefulness of actions by analyzing the game rules as opposed to the simulation results. Experimental results show the potential of this approach.

"Space-Consistent Game Equivalence Detection in General Game Playing," by Haifeng Zhang, Dangyi Liu, and Wenxin Li, discusses that GGP agents can efficiently enhance their intelligence by taking advantage of experience from past games. The authors argue that it is necessary for agents to detect equivalence between games. This paper defines game equivalence formally and concentrates on a specific scale, space-consistent game equivalence (SCGE). To detect SCGE, an approach is proposed mainly reducing the complex problem to some well-studied problems. An evaluation of the approach is performed at the end.

These proceedings would not have been produced without the help of many persons. In particular, we would like to mention the authors and reviewers for their help. Moreover, the organizers of IJCAI 2015 contributed substantially by bringing the researchers together.

March 2016   Tristan Cazenave
                                                                Mark H.M. Winands
                                                                 Stefan Edelkamp
                                                                 Stephan Schiffel
                                                               Michael Thielscher
                                                                  Julian Togelius

Organization

Program Chairs

Tristan Cazenave	Université Paris–Dauphine, France
Mark Winands	Maastricht University, The Netherlands
Stefan Edelkamp	University of Bremen, Germany
Stephan Schiffel	Reykjavik University, Iceland
Michael Thielscher	University of New South Wales, Australia
Julian Togelius	New York University, USA

Program Committee

Yngvi Björnsson	Reykjavik University, Iceland
Bruno Bouzy	Université Paris–Descartes, France
Tristan Cazenave	Université Paris–Dauphine, France
Stefan Edelkamp	University of Bremen, Germany
Michael Genesereth	Stanford University, USA
Hiroyuki Iida	Japan Advanced Institute of Science and Technology, Japan
Nicolas Jouandeau	Université Paris 8, France
Lukasz Kaiser	Université Paris–Diderot, France
Sylvain Lagrue	Université d'Artois, France
Marc Lanctot	Google DeepMind, UK
Simon Lucas	University of Essex, UK
Jacek Mańdziuk	Warsaw University of Technology, Poland
Jean Méhat	Université Paris 8, France
Martin Müller	University of Alberta, Canada
Diego Perez	University of Essex, UK
Thomas Philip Runarsson	University of Iceland, Iceland
Arpad Rimmel	Supelec, France
Ji Ruan	Auckland University of Technology, New Zealand
Abdallah Saffidine	University of New South Wales, Australia
Spyridon Samothrakis	University of Essex, UK
Tom Schaul	Google DeepMind, UK
Stephan Schiffel	Reykjavik University, Iceland
Sam Schreiber	Google Inc., USA
Nathan Sturtevant	University of Denver, USA
Fabien Teytaud	Université du Littoral Côte d'Opale, France
Michael Thielscher	University of New South Wales, Australia

Julian Togelius	New York University, USA
Mark Winands	Maastricht University, The Netherlands
Shi-Jim Yen	National Dong Hwa University, Taiwan

Additional Reviewers

Sumedh Ghaisas
Chiara Sironi
Maciej Świechowski

Contents

Computer Games Workshop 2015

Challenges and Progress on Using Large Lossy Endgame Databases in Chinese Checkers

Nathan R. Sturtevant[(✉)]

Department of Computer Science, University of Denver, Denver, CO, USA
sturtevant@cs.du.edu

Abstract. A common evaluation function for playing Chinese Checkers with two or more players has been the single-agent distance across the board. This is an abstraction of a perfect heuristic, because it ignores the interactions between the players in the game. Previous work has studied these heuristics for smaller versions of the game, including 6-piece data for a board with 49 locations and 81 locations which have 13.98 million and 324.5 million combinations respectively. The single-agent solution to the full game of Chinese Checkers has 81 locations and 10 pieces per player. This results in 1.88 trillion possible positions and is stored using 500 GB of disk space. In this paper we report results from a preliminary study on how to best use the data to improve the play of a Chinese Checkers program.

1 Introduction

Endgame databases have been a key component of game-playing programs in games like Checkers [23] and Chess [15], where they contain precise information that is easier to pre-compute than it is to summarize in heuristics or machine-tuned evaluation functions. Endgame databases are most useful when a game converges to simpler positions that may take a long time to resolve. In this case, the size of the endgame databases are small relative to the size of the game, and the computation associated with the endgame database would be non-trivial to reproduce at runtime. For instance, work on 7-piece chess endgame databases has discovered a position where 549 moves are required for mate [1]. These lines of play would not be discovered by programs that just attempt to find the best move from the same position, as the depth of search required is far beyond what could be found by normal time controls during a competitive game. Similarly, work on Checkers databases helped prove that a well-studied position from 1800, once assumed to be a win, was actually a draw [22]. Databases have also been successfully built for games like Chinese Chess [7].

Endgame databases are not universally useful, however, particularly when there are many possible endgame positions and when the resolution of each position is simple. In games such as card games, for example, endgame databases are essentially useless. There are $\frac{52!}{44!2^4} = 1.9$ trillion combinations of two cards that 4 players can have at the end of a trick-based card game (assuming a standard 52 card deck), but there are at most 16 ways to play out each possible hand.

© Springer International Publishing Switzerland 2016
T. Cazenave et al. (Eds.): CGW 2015/GIGA 2015, CCIS 614, pp. 3–15, 2016.
DOI: 10.1007/978-3-319-39402-2_1

In this case it will likely be faster to compute the result at runtime than to look the result up in the endgame database.

The game of Chinese Checkers is unique because the game decomposes into a single-agent game as it nears completion. This means that a single-agent heuristic can be used as an endgame database. That is, we can solve the single-agent version of the game to provide perfect distances to the end of the game once players' pieces have separated. But, there is also significant information about piece formation encoded in a full single-agent solution that is helpful for playing the game before it decomposes into a single agent game [25].

Work in the game of Chinese Checkers has often used a smaller board containing 49 locations, instead of the more common 81 location board, in order to facilitate the use of the single-agent solution both as a heuristic during search and as an endgame database [16,20,25]. We recently co-authored a study [18] looking at larger games, in particular the 81 location board with 6 pieces. But, nobody has studied the use of endgame databases in the full game with 81 locations and 10 pieces per player.

We recently built the single-agent solution [24] with 1.88 trillion positions, and requiring 500 GB of disk storage, far more than is found on typical machines. Furthermore, to save space, the data is stored modulo 15, using 4 bits per state. This means that additional work must be done to accurately recover the true distance of any given state. If there are errors in the recovery process, queries to the database may occasionally return incorrect values.

The goal of this paper is to study initial approaches for using this data to improve performance in Chinese Checkers. We use the data to create players based on minimax and MCTS search, comparing performance and uses of the data. We find that accuracy in retrieving values from the database is a primary concern, and that more work is needed to ensure accuracy of the database values. Preliminary experimental results suggest that minimax-based players perform better than MCTS-based players with using large endgame databases.

2   Background

In this section we cover a few important background details for this paper, including information about the game of Chinese Checkers and the algorithms that we will use to play the game.

2.1   Chinese Checkers

Chinese Checkers is a game played by 2–6 players on a star-shaped board shown in Fig. 1(a). In the two-player version of the game the players start with their pieces on the top and bottom of the board. The goal is for the players to get their pieces to the other side of the board. Legal movement is shown in Fig. 1(b). Pieces can either move in steps to one of the 6 adjacent locations, such as from location (c) to (d), or they can move in jumps, as shown by the move from (e) to (f). Jumps can only be taken when a piece is adjacent to another piece, and the location opposite to the adjacent piece is free; jumps can be taken over any piece regardless of ownership. Jumps can also be chained together, so the piece at (e) can move directly to (g) by taking two jumps as part of the same turn. Note that a variation on these jumping rules is sometimes played where pieces can jump more than one space at a time across the board [28].

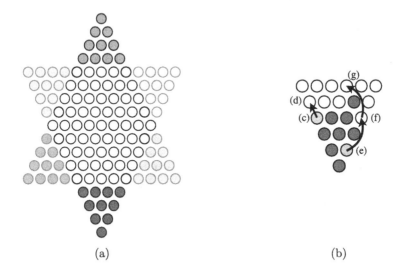

(a)   (b)

Fig. 1. Chinese Checkers board (a) and legal moves (b).

Most rules that we have seen for the game are ambiguous about some options for play, so we have introduced several conventions that simplify the game for computer play. First, we consider the game to be won by a player if a player's goal area is filled with pieces and at least one of them belongs to that player. This prevents a player from blocking their opponent's goal to prevent a win. Furthermore, if a state in the game is repeated, we consider this a loss by repetition for the player that moved into the repeated state.

Single Agent Chinese Checkers: The single-agent version of Chinese Checkers has $\frac{81!}{10!71!} = 1,878,392,407,320$ possible states. The Chinese Checkers board is symmetric around the x (horizontal) axis, so we can use the same single-agent solution for both players by flipping the board around the horizontally axis. The board is also symmetric around the y (vertical) axis, so the number of states which must be stored can be reduced. We use a quick, but slightly imperfect, scheme for computing symmetry: A state is not stored if (1) the first piece from the top of the board is on the right hand side of the board (excluding the center line) or (2) the first piece is on the center line of the board and the second piece is on the right hand side of the board. Using this scheme the number of stored states can be reduced to 1,072,763,999,648. (A perfect scheme would consider all pieces instead of just the first two; it would be slower to compute but use slightly less space.)

We use a write-minimizing breadth-first search [24] to generate and store the distance from each of these states to the goal state.[1] To save space and make the search more efficient we store the solution using only 4 bits per entry. This results in 16 values per entry, but one value is needed to mark unseen states during the search, so the resulting data stores the depth of each state modulo 15. As a result, states at depth 0 and 15 both have the same value in the database: 0. We discuss the consequences of this compression later in the paper.

In Fig. 2(a) we show, given the location of the first piece on the board, the size of the data needed to store the single-agent distances for all board configurations with a given first piece location. The locations are labeled in Fig. 2(b) - the top of the board is location 0, and the middle row starts with location 36.

The full data, with the first piece in position 0, requires 500 GB to store. The position in Fig. 2(c) has the first piece in position 1; if we only lookup positions with the first piece in location 1 or higher, we will only need 433 GB of storage. The board in Fig. 2(d) has the first piece in position 17. If we want to lookup all positions with the first piece in position 17 or higher, we need to store the data for positions 15 and higher, requiring 61.5 GB. This is due to the symmetry used for lookups: Our symmetry rules prevent us from storing positions where the first piece is in location 20. Instead, a symmetric lookup is used; flipping such a position around the y-axis moves this piece into location 15. Thus, we always store full rows of the single-agent data to ensure that symmetric lookups are always possible.

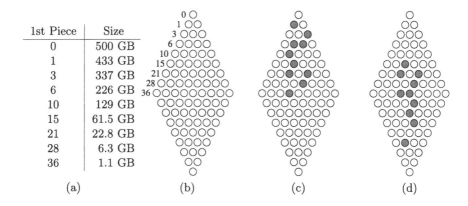

1st Piece	Size
0	500 GB
1	433 GB
3	337 GB
6	226 GB
10	129 GB
15	61.5 GB
21	22.8 GB
28	6.3 GB
36	1.1 GB

(a)   (b)   (c)   (d)

Fig. 2. The size of the single-agent data for various points on the board.

2.2 Minimax and Alpha-Beta Pruning

Minimax is the basic algorithm that was used to build expert-level programs for games like Chess [5] and Checkers [22]. It relies on an evaluation function that

---

[1] Due to symmetry either the start or the goal state can be used, although the choice influences whether subsets of data can be efficiently loaded.

maps states to numerical values which estimate the true value of a state. A large number of enhancements have been proposed to minimax search to improve performance [21].

We build a basic minimax program that is enhanced with iterative deepening, alpha-beta pruning, the history heuristic and transposition tables. The base evaluation function is just the distance of a player's pieces across the board, but we have also used $TD(\lambda)$ [27] with linear regression to learn an evaluation function. A complete description of the details of our approach is outside the scope of this paper, but it is worth noting that this is not the first use of learning in Chinese Checkers, although it is the first use of $TD(\lambda)$ that we could find in the literature. Samadi et al. report using learning in their program without further detail [19]. Hutton [11] trained genetic algorithms on training data for mid-game play.

2.3   Monte-Carlo Tree Search

Monte-Carlo Tree Search (MCTS) is an alternate to minimax which has been used successfully in a broad range of other games such as Go [9], Hex [10], Hearts [26], and Amazons [14]. MCTS algorithms build an in-memory tree and then sample the leaves of the tree (e.g. with random playouts to the end of the game) as an alternate to evaluating states with an evaluation function. The tree grows with each playout, and is non-uniformly biased towards better parts of the search space. In many cases a good evaluation function for moves is needed, to ensure that playouts are reasonable and finite. Similarly with minimax, there are numerous enhancements that have been proposed for MCTS [4]. But, unlike minimax enhancements, it is still an open question of which enhancements are redundant relative to each other.

The primary enhancement we use here is forward pruning [14] as well as dynamic early termination [3]. We also use epsilon-greedy playouts [26] and enhanced playout policies to improve play.

We have experimented with slowing the growth of the MCTS tree [18], with initializing states with offline values [8], but current results are inconclusive. Other approaches that would be worth considering in the future include implicit minimax backups [13] and progressive bias [6].

3   Lossy Chinese Checkers Endgame Database

As mentioned previously, the single-agent data for Chinese Checkers is stored modulo 15, but the maximum distance between any two states in the single-agent Chinese Checkers state space is 33 moves [24]. The majority of positions (99.21 %) in the game are between 15 and 29 moves from the goal, but since every game terminates with a position at depth 0, a significant portion of positions seen in practice will be between 0 and 14 moves from the goal.

This creates the need for a classifier that will take a state and the stored depth as input and return the true distance to the goal as output. A simple

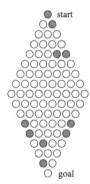

Fig. 3. Chinese Checkers board at depth 18 which must be classified.

lower bound on the number of moves to reach the goal can be computed by (1) the number of pieces that are not yet in the goal area and (2) the number of blank horizontal lines between the goal area and the farthest piece from the goal. This is related to observations made by Bell [2], modified for our purposes here. In particular, it is impossible for the furthest back piece on the board to skip rows when moving towards the goal in the single-agent game, since it either must step through each row between it and the goal, or it will jump over another piece en route to the goal.

Unfortunately, this estimate alone is not enough for a perfect predictor. Other factors such as non-adjacent pieces can be used to improve the classifier, but not with 100 % accuracy. We show a sample board that must be classified in Fig. 3. In this board position there are 8 pieces outside of the goal area. Furthermore, there are 8 empty rows between the goal and the furthest removed piece. Thus, no solution can exist which does not take at least 16 moves to reach the goal. The single-agent solution data says this position is at depth 3. Given the lower bound, the state is either 18 or 33 moves from the goal. In this case, the lower bound is nearly perfect. Our classifier currently mis-classifies this position, thinking that it will take more work to get the pieces split on either side of the board (just outside the goal) into the goal area.

To measure the accuracy of our predictor we sampled every 100,000 states (approximately) for the first 64.112 % of the full state space. We looked at the predicted depth of each state as well as the predicted depth of its neighbors to determine whether the prediction accuracy is correct. If a state and one or more of its neighbors differ by more than 1 step, the predictor must be incorrect. Overall, we looked at approximately 10 million states and their neighbors. When there is an error, on average 3.8 neighbors of the same parent show the error. Overall, 1 in 1800 parents has a child with a heuristic error. This is very small as a percentage, but in a search that expands millions of nodes, it is very likely that the search will encounter nodes with incorrect depth values.

4   Combining with Alpha-Beta Search

The first player we create is a traditional player using minimax with alpha-beta pruning [17]. This player is enhanced using iterative deepening, transposition tables and the history heuristic [21]. Moves are partially randomized to avoid identical play on repeated games. Only forward moves are allowed, as this significantly improves performance over considering all possible moves.

We implement two evaluation functions for this player. In the first each player just attempts to minimize the number of rows between their pieces and the goal. The second evaluation function is trained with self-play using linear regression and $TD(\lambda)$. This evaluation function learns a weight for each location on the board for each player; it is strictly more expressive than the simple distance evaluation, since the distance evaluation function can be implemented by weighting each location according to its distance to the goal. This approach learns a total of 162 weights, one for each position on the board for each player.

We then integrate the single-agent data as an endgame database into this code base. It is important to note a key difference between endgame databases here and in traditional programs. Because we can use the single-agent data both as a heuristic and as an endgame database, we should not terminate the search immediately once a lookup is available. If the players' pieces are not adequately separated, the single-agent data is only a heuristic. This data gets more accurate the closer we get to the end of the game. Thus, using it at the end of a deeper search will improve performance. In practice, we don't have to modify our program to use the exact values when the game decomposes; it does not hurt the minimax search to lookahead further than the beginning of the endgame database, because the result of the minimax computation stays the same.

4.1   Experimental Results

We experiment primarily with the 1.1 GB database beginning at location 36 and the 61.5 GB database beginning at location 15. We have a 16-core 2.6 GHz AMD Opteron server with 64 GB of RAM, so this is the largest database we can load into memory. At leaves of the tree, if both players are able to use the database lookup, then difference in distances for each player is used as the evaluation function. If only one or neither player is able to use the database lookup, then both players use the regular evaluation function.

We present the results in Table 1. M indicates that the player is using the minimax algorithm to play. We use subscripts to indicate whether the database is being used as part of the evaluation function. The number used as the subscript is the first piece in the loaded database. Superscripts are used to indicate the evaluation function that is used. T for the trained evaluation function, and D for the distance-based evaluation function. So, the player labeled M_{36}^{T} is using minimax and a trained evaluation function in conjunction with the endgame database starting at location 36 on the board.

Each set of players is matched up 200 times - 100 times as the first player and 100 times as the second player. In experimental results we refer to players as Player 1 and Player 2 only for the simplicity of distinguishing player types and win rates, not for indicating which player went first. We record and report the win/loss percentage for each player. One player is better than another with 95 % confidence if it wins 114 games (57 %); 118 games (59 %) are needed for 99 % confidence. Players are allowed 1 s per move. While the exact search depth depends on the player, the average search depth is over 6 ply in practice. All further experiments use this same setup.

Table 1. Minimax win rates.

Player 1	Player 2	P1 Win %	P2 Win %
M^T	M^D	76.0 %	24.0 %
M_{36}^T	M^T	63.0 %	37.0 %
M_{36}^D	M^D	67.5 %	32.5 %
M_{36}^T	M_{36}^D	80.0 %	20.0 %
M_{15}^D	M^D	18.5 %	81.5 %
M_{15}^D*	M^D	52.5 %	47.5 %
M_{15}^D*	M_{36}^D	62.5 %	37.5 %

In the first half of the table we compare results using the database starting at piece location 36. The trained player soundly beats the regular distance player[2], both players are significantly stronger with the database than without, and the trained player beats the distance player by a significant margin when using the database. However, when the larger database is used (M_{15}^D vs M^D), the performance drops significantly. We hypothesized that this was the result of the errors in computing exact distances in the database: The games that are won by this player exhibit strong play. In games that are lost, however, the player with the database exhibits pathological play: the program moves to a seemingly random configuration of the board and then does its best to preserve that configuration. This is clearly a result of imperfect recovery of distances from the database.

To measure this, we implemented an alternate version of minimax that tracks the database values during play by looking the distance up from disk after every move by each player. Since the change in the optimal number of moves to the goal between a parent and child state (i.e. from applying a single action) cannot be large,[3] we use this to maintain more accurate distance estimates. The single-agent distance of the current board configuration for each player is updated incrementally through the game, starting at the optimal single-agent distance of 27.

---

[2] In our original implementation the algorithms were indistinguishable. Improving the efficiency of our TD learning, by taking advantage of the binary features, significantly improved the performance of the TD player.

[3] It can change by more than one because we may jump over an opponent's piece, which is not accounted for in the single-agent data.

We assume that the distance from two adjacent states to the goal will never increase by more than 5 or decrease by less than 10.

It is too expensive to perform a disk access for every node in the search tree, but once the start position for each player is within the in-memory database, these distances can also be incrementally maintained within the minimax tree during search. Thus, at the leaves of the tree, accurate distances can be used. The player that uses this improved distance estimation is in the last two line of Table 1 - M_{15}^D*. When played against a regular minimax player, this program is not significantly stronger than not using the databases. However, when played against a player using the small endgame databases, the program wins 62.5 % of games. Analysis of the games indicates that there are likely still small errors in the database distance lookups, as we have found positions where the program does not take clearly winning moves. But, this data point clearly indicates that errors in retrieving accurate endgame data are the reason for the poor play, since improving this accuracy significantly improves play.

We are working to build better classifiers that will improve the accuracy in estimating true distances from the database given the modulo distance.

5   Combining with MCTS

Next we integrate the endgame data into a MCTS approach. Our MCTS player is based on UCT [12]. We use the UCB1 rule to select the best node in the tree for exploration. Sampling outside of the tree is done using epsilon-greedy playouts [26]. Playouts are cutoff at depth 10, where a static evaluation function is used to evaluate the state [14]. (Experiments with different cutoff depths indicated that this depth produced robust performance across other parameter settings.) Like in our minimax program, we restrict the program to only taking forward moves. Our playouts are also biased – unless the epsilon-greedy rule takes precedence ($\epsilon = 0.05$), the largest forward move is always taken during each playout step.

We first began by replacing the heuristic evaluation at depth 10 with the value from the database, assuming that both players' lookups were available in the database. Although this basic approach seemed like it should improve performance, we needed to make a few changes to make this successful. First, we immediately cut off a playout once both players' pieces have separated so that a perfect evaluation is possible. Second, we use the database for the evaluation function if it is available for both players, otherwise we use the basic heuristic evaluation. Third, we must be more careful to account for the player to move when performing evaluations, since the player to move at the leaves of the tree is non-uniform. If there is a tie in distances to the goal, the player with the first move will win.

This first rule is particular important. Without it, our program plays poorly despite a perfect evaluation function. It appears that this occurs primarily because the random playouts corrupt the perfect leaf values and because it takes some time for the MCTS tree to converge its playout values to the perfect values after every move is sampled. While the need to immediately use the perfect values once they are available seems obvious, this is not needed in the minimax program. Similarly, because all leaf nodes are at the same depth, there is no need to account for the player to move in the evaluation function of a minimax player, since the bonus would be added equally to all nodes.

Besides the approach just described, there are several other ways to use the endgame databases in MCTS: Similar to our previous analysis [18], performing random playouts until both players separate to get a more accurate evaluation function did not work well in practice. Also, using the data directly from disk is too slow and results in very poor performance, even though we have the data on a SSD for fast access. We experimented with other playout depths, and distances in the range 10–20 provided the best performance. Slowing the growth of the MCTS tree by growing the tree more slowly as done in our previous work [18] did not have a significant impact on performance.

5.1   Experimental Results

We now compare the performance of MCTS (UCT) implementations using the single-agent database as an evaluation function. Results are in Table 2. U designates a player using UCT. As before, the subscript indicates the first piece in the endgame database that is used by the player during play. In this experiment all players use the distance-based evaluation as the base evaluation function. As before, 1 s is allowed for each move.

Table 2. MCTS/UCT win rates.

Player 1	Player 2	P1 Win %	P2 Win %
U_{36}^D	U^D	49.0 %	51.0 %
U_{28}^D	U_{36}^D	57.0 %	43.0 %
U_{15}^D	U^D	74.5 %	25.5 %
U_{15}^D	U_{36}^D	63.5 %	36.5 %

In the first line we see that adding the endgame database starting at position 36 does not significantly improve or degrade performance over the basic player. But, using the endgame database starting at piece 28 wins over the endgame database starting at piece 36. Similarly, using the endgame database starting at 15 wins significantly over just using the distance evaluation and over the database starting at piece 36. We did not compare the database starting at location 15 with the database starting at location 28 due to memory constraints, although sharing the database between players would allow this comparison.

Our results suggest that MCTS does a better job than minimax of tolerating errors in the single-agent data, as performance improves as we use the larger databases, even without the correction procedures needed for minimax. However, we need to be more careful about when and how we use the endgame data in MCTS.

6   Comparing Minimax and MCTS

Given that we have successfully improved the performance of our two independent approaches using endgame databases, we now compare the performance across techniques. We begin with a baseline comparison without using endgame databases, shown in the top two lines of Table 3.

Table 3. Minimax versus UCT.

Player 1	Player 2	P1 Win %	P2 Win %
U^D	M^D	77.5 %	22.5 %
U^D	M^T	63.0 %	37.0 %
M^D_{36}	U^D_{36}	60.5 %	39.5 %
M^D_{36}	U^D_{15}	90.5 %	9.5 %

Without the endgame databases, UCT beats both the regular and trained minimax player by a significant margin. But, as shown in the bottom half of the table, when minimax is given the database, it outperforms UCT by a significant margin.

Furthermore, the UCT player with the database starting at location 15 does worse than the player with the database at location 36 against minimax. This result suggests that the UCT player does well against other UCT players, but does not necessarily do well against other player types. We expect that improving the accuracy of the UCT endgame estimates could further improve performance.

But, we also note that one of the strengths of UCT is that it gets long-term strategic information about a game from its playouts. Adding endgame databases seems to duplicate this strength. Minimax, on the other hand, does well in local tactics because of the full search, but misses out on longer-term strategies. Thus, endgame databases seem to have the potential to complement the performance of minimax more than UCT.

Further experiments and implementations will be needed to understand this difference in performance more deeply and to clearly isolate the factors that influence performance with endgame databases.

We note that these results are different than in our preliminary study, where minimax performed worse that UCT [18]. There are several notable differences between this work. First, our minimax implementation here is much stronger than in previous work. Second, we are using more pieces on the board, where there is more congestion. Third, we are using better time controls for our experiments, using only time instead of node counts. Finally, we have error in our database lookups in this work, while our previous work had no error.

7   Conclusions and Future Work

This paper describes the first work in using large endgame databases in Chinese Checkers. Several challenges are faced in how to properly integrate the endgame databases and how to correct errors that occur because of the compression that is used when storing the endgame data. Currently, minimax-based approaches are able to do a better job of using the endgame data than MCTS approaches, but with further study this could still change.

The addition of opening books and further search enhancements could improve either player, particularly since the programs do not take the moves that advance their pieces most quickly across the board at the beginning of the game. (The endgame database can be used for this purpose as an opening book as well.)

Improving our classifier is also an important step to improving performance. Currently our classifier uses relatively simple rules to estimate the true distance to the goal given the modulo distance. Using linear or logistic regression to train a classifier could result in better performance. We should also be able to enhance our UCT player to improve its own estimates during playouts.

Another important step is to measure the robustness of both minimax and MCTS to errors in the evaluation function to see which approach is more tolerant to the type of errors that occur in our evaluation function.

Finally, we need to understand more deeply why our UCT player is able to beat other UCT players by a significant margin, but not minimax players. This is related to the quality of our classifier, but also to a fundamental understanding of the strengths of each approach.

Acknowledgments. This paper benefited from research by a summer student, Evan Boucher, who worked on the problem of determining the true distance of a state from the goal given the modulo distance.

References

1. 8 longest 7-man checkmates. http://tb7.chessok.com/articles/Top8DTM_eng. Accessed 11 May 2015
2. Bell, G.I.: The shortest game of Chinese Checkers and related problems. CoRR abs/0803.1245 (2008). http://arxiv.org/abs/0803.1245
3. Bouzy, B.: Old-fashioned computer Go vs Monte-Carlo Go. In: IEEE Symposium on Computational Intelligence in Games (CIG) (2007). Invited Tutorial
4. Browne, C., Powley, E.J., Whitehouse, D., Lucas, S.M., Cowling, P.I., Rohlfshagen, P., Tavener, S., Perez, D., Samothrakis, S., Colton, S.: A survey of monte carlo tree search methods. IEEE Trans. Comput. Intell. AI Games **4**(1), 1–43 (2012)
5. Campbell, M., Hoane Jr., A.J., Hsu, F.: Deep blue. Artif. Intell. **134**(1–2), 57–83 (2002)
6. Chaslot, G.M.J.B., Winands, M.H.M., van den Herik, H.J., Uiterwijk, J.W.H.M., Bouzy, B.: Progressive strategies for Monte-Carlo tree search. New Math. Nat. Comput. **4**(3), 343–357 (2008)
7. Fang, H., Hsu, T., Hsu, S.-C.: Construction of Chinese chess endgame databases by retrograde analysis. In: Marsland, T., Frank, I. (eds.) CG 2001. LNCS, vol. 2063, pp. 96–114. Springer, Heidelberg (2002)
8. Gelly, S., Silver, D.: Combining online and offline knowledge in UCT. In: Ghahramani, Z. (ed.) Machine Learning. ACM International Conference Proceeding Series, vol. 227, pp. 273–280. ACM, New York (2007)

9. Gelly, S., Silver, D.: Achieving master level play in 9 x 9 computer go. In: Fox, D., Gomes, C.P. (eds.) AAAI, pp. 1537–1540. AAAI Press, Menlo Park (2008)
10. Huang, S.-C., Arneson, B., Hayward, R.B., Müller, M., Pawlewicz, J.: MoHex 2.0: a pattern-based MCTS hex player. In: van den Herik, H.J., Iida, H., Plaat, A. (eds.) CG 2013. LNCS, vol. 8427, pp. 60–71. Springer, Heidelberg (2014)
11. Hutton, A.: Developing Computer Opponents for Chinese Checkers. Master's thesis, University of Glasgow (2001)
12. Kocsis, L., Szepesvári, C.: Bandit based Monte-Carlo planning. In: Fürnkranz, J., Scheffer, T., Spiliopoulou, M. (eds.) ECML 2006. LNCS (LNAI), vol. 4212, pp. 282–293. Springer, Heidelberg (2006)
13. Lanctot, M., Winands, M.H.M., Pepels, T., Sturtevant, N.R.: Monte Carlo tree search with heuristic evaluations using implicit minimax backups. In: 2014 IEEE Conference on Computational Intelligence and Games, CIG 2014, Dortmund, Germany, 26–29 August 2014, pp. 341–348. IEEE (2014)
14. Lorentz, R.J.: Amazons discover Monte-Carlo. In: van den Herik, H.J., Xu, X., Ma, Z., Winands, M.H.M. (eds.) CG 2008. LNCS, vol. 5131, pp. 13–24. Springer, Heidelberg (2008)
15. Nalimov, E., Haworth, G.M., Heinz, E.A.: Space-efficient indexing of endgame tables for chess. ICGA J. **23**(3), 148–162 (2000)
16. Nijssen, J.P.A.M., Winands, M.H.M.: Enhancements for multi-player Monte-Carlo tree search. In: van den Herik, H.J., Iida, H., Plaat, A. (eds.) CG 2010. LNCS, vol. 6515, pp. 238–249. Springer, Heidelberg (2011)
17. Pearl, J.: The solution for the branching factor of the alpha-beta pruning algorithm and its optimality. Commun. ACM **25**(8), 559–564 (1982)
18. Roschke, M., Sturtevant, N.R.: UCT enhancements in Chinese Checkers using an endgame database. In: Cazenave, T., Winands, M.H.M., Iida, H. (eds.) Computer Games (CGW 2013). CCIS, vol. 408, pp. 57–70. Springer International Publishing, Switzerland (2014)
19. Samadi, M., Asr, F.T., Schaeffer, J., Azimifar, Z.: Extending the applicability of pattern and endgame databases. IEEE Trans. Comput. Intell. AI Games **1**(1), 28–38 (2009)
20. Samadi, M., Schaeffer, J., Asr, F.T., Samar, M., Azimifar, Z.: Using abstraction in two-player games. In: ECAI, pp. 545–549 (2008)
21. Schaeffer, J.: The history heuristic and alpha-beta search enhancements in practice. IEEE Trans. Pattern Anal. Mach. Intell. **11**(11), 1203–1212 (1989)
22. Schaeffer, J.: One Jump Ahead - Challenging Human Supremacy in Checkers. Springer, New York (1997)
23. Schaeffer, J., Björnsson, Y., Burch, N., Lake, R., Lu, P., Sutphen, S.: Building the checkers 10-piece endgame databases. Adv. Comput. Games **10**, 193–210 (2003)
24. Sturtevant, N.R., Rutherford, M.J.: Minimizing writes in parallel external memory search. In: International Joint Conference on Artificial Intelligence (IJCAI) (2013)
25. Sturtevant, N.R.: A comparison of algorithms for multi-player games. In: Schaeffer, J., Müller, M., Björnsson, Y. (eds.) CG 2002. LNCS, vol. 2883, pp. 108–122. Springer, Heidelberg (2003)
26. Sturtevant, N.R.: An analysis of UCT in multi-player games. ICGA J. **31**(4), 195–208 (2008)
27. Sutton, R.S., Barto, A.G.: Reinforcement Learning: An Introduction. MIT Press, Cambridge (1998)
28. Tong, K.B.: Intelligent Strategy for Two-person Non-random Perfect Information Zero-sum Game. Master's thesis, Chinese University of Hong Kong (2003)

Sequential Halving for Partially Observable Games

Tom Pepels[1]([⊠]), Tristan Cazenave[2], and Mark H.M. Winands[1]

[1] Department of Data Science and Knowledge Engineering,
Maastricht University, Maastricht, The Netherlands
{tom.pepels,m.winands}@maastrichtuniversity.nl
[2] LAMSADE - Université Paris-Dauphine, Paris, France
cazenave@lamsade.dauphine.fr

Abstract. This paper investigates Sequential Halving as a selection policy in the following four partially observable games: Go Fish, Lost Cities, Phantom Domineering, and Phantom Go. Additionally, H-MCTS is studied, which uses Sequential Halving at the root of the search tree, and UCB elsewhere. Experimental results reveal that H-MCTS performs the best in Go Fish, whereas its performance is on par in Lost Cities and Phantom Domineering. Sequential Halving as a flat Monte-Carlo Search appears to be the stronger technique in Phantom Go.

1 Introduction

Partially observable games introduce the complexity of uncertainty in gameplay. In partially observable games, some element of the game is not directly observable. The unknown element can be introduced by hiding certain parts of the current state to the player (e.g., hiding the rank of piece in Stratego), in game theory this is also called imperfect information. Other than in fully observable games, we cannot directly search for sequences of actions leading to promising moves using the partially visible state. In this paper we discuss four different partially observable games: Go Fish and Lost Cities, which are card games with imperfect information, and the so-called phantom games: Phantom Domineering and Phantom Go.

Different approaches have been suggested for handling partial observability in Monte-Carlo Tree Search (MCTS) in such domains. Such as Determinized UCT [13] where a random game state is sampled before the search (*i.e.,* determinized), and multiple trees are maintained per determinization. The recently introduced Information Set MCTS [13] maintains information sets of states reachable in the current determinization in the tree, as such re-using statistics over multiple determinizations in the tree.

In this paper we investigate the effects of using Sequential Halving [16] as a selection policy in partially observable games. We continue to study the Hybrid MCTS [20] algorithm, introduced as a method of minimizing simple and cumulative regret simultaneously during search.

© Springer International Publishing Switzerland 2016
T. Cazenave et al. (Eds.): CGW 2015/GIGA 2015, CCIS 614, pp. 16–29, 2016.
DOI: 10.1007/978-3-319-39402-2_2

The paper is structured as follows. First, in Sect. 2, we give a brief overview of MCTS. Next, in Sect. 3 we discuss Sequential Halving, and how it may be applied to MCTS in partially observable games. After this we describe the test domains in Sect. 4. Finally, we show the experimental results in Sect. 5, and discuss our conclusions and directions for future research in Sects. 6 and 7.

2   Monte-Carlo Tree Search

Monte-Carlo Tree Search (MCTS) is a best-first search method based on random sampling by Monte-Carlo simulations of the state space of a domain [12,17]. In game play, this means that decisions are made based on the results of randomly simulated play-outs. MCTS has been successfully applied to various turn-based games such as Go [22], Lines of Action [26], and Hex [1]. Moreover, MCTS has been used for agents playing real-time games such as the Physical Traveling Salesman [21], real-time strategy games [4], and Ms Pac-Man [19], but also in real-life domains such as optimization, scheduling, and security [6].

In MCTS, a tree is built incrementally over time, which maintains statistics at each node corresponding to the rewards collected at those nodes and number of times they have been visited. The root of this tree corresponds to the current position. MCTS performs iteratively simulations until a computational threshold is reached, *i.e.*, a set number of simulations, an upper limit on memory usage, or a time constraint.

Each MCTS simulation consists of two main steps, (1) the *selection* step, where moves are selected and played inside the tree according to the selection policy until a leaf is *expanded*, and (2) the *play-out*, in which moves are played according to a simulation policy, outside the tree. At the end of each play-out a terminal state is reached and the result is *back-propagated* along the selected path in the tree from the expanded leaf to the root.

2.1   UCT

During the selection step, a policy is required to explore the tree to decide on promising options. For this reason, the Upper Confidence Bound applied to Trees (UCT) [17] was derived from the UCB1 [3] policy. In UCT, each node is treated as a multi-armed bandit problem whose arms are the moves that lead to different child nodes. UCT balances the exploitation of rewarding nodes whilst allowing exploration of lesser visited nodes. Consider a node p with children $I(p)$, then the policy determining which child i to select is defined as:

$$i^* = argmax_{i \in I(p)} \left\{ v_i + C\sqrt{\frac{\ln n_p}{n_i}} \right\}, \tag{1}$$

where v_i is the score of the child i based on the average result of simulations that visited it, n_p and n_i are the visit counts of the current node and its child, respectively. C is the exploration constant to tune. UCT is applied when the visit count of p is above a threshold T, otherwise a child is selected at random. UCB1 and consequently, UCT incorporate both exploitation and exploration.

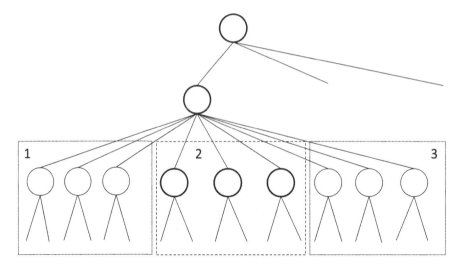

Fig. 1. Example of three determinizations within a single tree. The selected determinization is 2. All unreachable nodes in determinization 2 will not be selected.

2.2 MCTS in Partially Observable Games

To deal with games having imperfect information, determinization can be applied in the MCTS engine. The principle behind determinization is that, at the start of each simulation at the root, the hidden information is 'filled in', ensuring it is consistent with the history of the current match.

Determinization has been called "averaging over clairvoyance" [23], where players never try to hide or gain information, because in each determinization, all information is already available. Despite these shortcomings, it has produced strong results in the past, for instance in Monte-Carlo engines for the trick-based card game Bridge [15], the card game Skat [8], Scrabble [24], and Phantom Go [9].

Determinization in the MCTS framework has been applied in games such as Scotland Yard [18] and Lord of the Rings: The Confrontation [13]. It works as follows. For each MCTS simulation starting at the root the missing information is filled in a random manner. The determinization is used throughout the whole simulation. Next, there are two approaches to build and traverse the search tree.

The first approach is by generating a separate tree for each determinization [13]. After selecting a determinization at the root node, the corresponding tree is traversed. Based on majority voting [18] the final move can be selected. Each candidate move receives one vote from each tree where it is the move that was played most often. The candidate move with the highest number of votes is selected as the best move. If more moves are tied, the move with the highest number of visits over all trees is selected. The concept of separate-tree determinization is similar to root parallelization.

The second approach is using single-tree determinization [11,13,18]. When generating the tree, all possible moves from all possible determinizations are generated. When traversing the tree, only the moves consistent with the current determinization are considered. An example is given in Fig. 1. The advantage of this technique is that information is shared between different determinizations, increasing the amount of usable information. This type of determinization is also named Single-Observer Information Set Monte-Carlo Tree Search [13].

3   Sequential Halving and MCTS in Partially Observable Games

In this section we describe our approach to applying Hybrid MCTS [20] (H-MCTS) to partially observable games. H-MCTS is based on the concept of minimizing *simple regret* near the root, and *cumulative regret* in the rest of the tree. Simple regret is defined as the regret of not *recommending* the optimal move. Whereas cumulative regret is the sum over the regret of having *selected* suboptimal moves during sampling.

In their analysis of the links between simple and cumulative regret in multi-armed bandits, Bubeck *et al.* [7] found that upper bounds on cumulative regret lead to lower bounds on simple regret, and that the smaller the upper bound on the cumulative regret, the higher the lower bound on simple regret, regardless of the recommendation policy, *i.e.,* the smaller the cumulative regret, the larger the simple regret. As such, no policy can give an optimal guarantee on both simple and cumulative regret at the same time. Since UCB gives an optimal upper bound on cumulative regret, it cannot also provide optimal lower bounds on simple regret. Therefore, a combination of different regret minimizing selection methods in the same tree is used in H-MCTS.

This section is structured as follows, first we discuss Sequential Halving, a novel simple regret minimizing algorithm for multi-armed bandits, in Subsect. 3.1. Next, in Subsect. 3.2 we discuss how a hybrid search technique may be used in partially observable games.

3.1   Sequential Halving

Non-exploiting selection policies have been proposed to decrease simple regret at high rates in multi-armed bandits. Given that UCB1 [3] has an optimal rate of cumulative regret convergence, and the conflicting limits on the bounds on the regret types shown in [7], policies that have a higher rate of exploration than UCB1 are expected to have better bounds on simple regret. Sequential Halving (SH) [16] is a novel, pure exploration technique developed for minimizing simple regret in the multi-armed bandit (MAB) problem.

In many problems there are only one or two good decisions to be identified, this means that when using a pure exploration technique, a potentially large portion of the allocated budget is spent sampling suboptimal arms. Therefore, an efficient policy is required to ensure that inferior arms are not selected as often as arms with a high reward. Successive Rejects [2] was the first algorithm to show a high rate of decrease in simple regret. It works by dividing the total computational budget into distinct rounds. After each round, the single worst

---

Algorithm 1. Sequential Halving [16].

Input: total budget T, K arms
Output: recommendation J_T

1 $S_0 \leftarrow \{1, \ldots, K\}$, $B \leftarrow \lceil \log_2 K \rceil - 1$

2 **for** $k=0$ **to** B **do**

3   sample each arm $i \in S_k$, $n_k = \left\lfloor \dfrac{T}{|S_k| \lceil \log_2 |S| \rceil} \right\rfloor$ times

4   update the average reward of each arm based on the rewards

5      $S_{k+1} \leftarrow$ the $\lceil |S_k|/2 \rceil$ arms from S_k with the best average

6 **return** *the single element of* S_B

---

arm is removed from selection, and the algorithm is continued on the reduced subset of arms. Sequential Halving [16], was later introduced as an alternative to Successive Rejects, offering better performance in large-scale MAB problems.

SH divides search time into distinct rounds, during each of which, arms are sampled uniformly. After each round, the empirically worst half of the remaining arms are removed until a single one remains. The rounds are equally distributed such that each round is allocated approximately the same number of trials (budget), but with smaller subset of available arms to sample. SH is detailed in Algorithm 1.

3.2 Hybrid MCTS for Partially Observable Games

Hybrid MCTS (H-MCTS) has been proposed by Pepels *et al.* in [20]. The technique uses recursive Sequential Halving, or SHOT [10] to minimize simple regret near the root as depicted in Fig. 2. The hybrid technique has shown to improve performance in several domains, including Amazons, Atari Go and Breakthrough. Previous algorithms that use MCTS with simple regret minimizing selection methods showed similar improvements in recommended moves in Markov Decision Processes [14,25].

In this paper we apply H-MCTS to partially observable games. The problem with these domains is that, when using multiple determinizations during search, revisiting nodes may result in different playable moves. This is not a problem when using selection methods such as UCT, which are greedy and select moves based on the current statistics. However, because SH is a uniform exploration method, in order to guarantee its lower bound on simple regret it must be able to revisit the same node a predetermined number of times. In other words, available moves should not change in between visits of the algorithm, or its specifically designed budget allocation is no longer valid.

In all partially observable games, the current player always has knowledge over the current set of moves that he can play given a fully observable determinization. Therefore, at the root of the search tree, moves are consistent between visits. As such, SH can be used to uniformly explore moves at the root without problems. When using multiple determinizations in a single tree, as in IS-MCTS, however, it is no longer possible to use SH deeper in the tree.

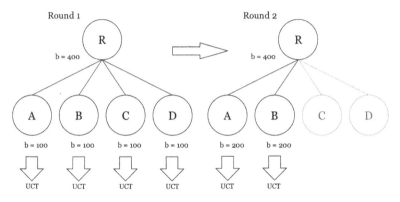

Fig. 2. Example rounds of H-MCTS with a budget limit $B = 150$. Sequential Halving is applied only at the root. On all other plies, UCT in the form of IS-MCTS is applied.

Each time a node is visited it may have a different subset of children based on the determinization (as depicted in Fig. 1). However, when using determinized UCT with a finite set of individual trees per determinization, SH can be used to select nodes deeper than the root, such an investigation is a possible direction for future research.

The approach is detailed in Algorithm 2. At the root, budget is allocated according to SH. For each sample, the appropriate IS-MCTS implementation can be used [13]. For this paper, based on the test domains (Sect. 4), we use single observer IS-MCTS.

4   Test Domains

In this section we discuss the partially observable games which are used in the experiments in Sect. 5. First, we describe the two card games: Go Fish and Lost Cities. Next, the phantom games Phantom Domineering and Phantom Go are explained.

Algorithm 2. Sequential Halving and Information Set MCTS.

Input: total budget T, K moves
Output: recommendation J_T

1 $S_0 \leftarrow \{1, \dots, K\}$, $B \leftarrow \lceil \log_2 K \rceil - 1$

2 **for** $k=0$ **to** B **do**

3   **for** *each move* $i \in S_k$ **do**

4          $n_k \leftarrow \left\lfloor \dfrac{T}{|S_k| \lceil \log_2 |S| \rceil} \right\rfloor$

5   **for** $n=0$ **to** n_k **do**

6   select a new determinization d at random

7   sample move i using IS-MCTS and determinization d

8   update the average of i reward based on the sample

9      $S_{k+1} \leftarrow$ the $\lceil |S_k|/2 \rceil$ moves from S_k with the best average

10 **return** *the single element of* S_B

4.1   Card Games

In both Go Fish and Lost Cities, cards are drawn from a randomly shuffled deck, limiting the possible predictions of future states. Moreover, in both games, moves available to the opponent are either partially or completely invisible. However, whenever a move is made, it becomes immediately known to both players. As these games progress, more information regarding the actual game state becomes available to both players.

Go Fish is a card game which is generally played by multiple players. The goal is to collect as many 'books' of 4 cards of equal rank. All players hide their cards from each other, and only finished books of four cards are placed face-up on the table. Each turn, a player may ask a single other player for a specific rank. If the questioned player has any cards of the requested rank in his hand, he gives them to the requesting player, which may consequently make a new request. If the questioned player does not possess a card of the requested rank, the questioning player must 'go fish', drawing a card from the stack, and the turn moves to the next player. The game ends when there are no more cards on the stack, and the player with the most finished books wins the game.

In our implementation, the game was slightly modified to allow it to be played by two players. Both players receive seven cards in hand at the start of the game. Moreover, the finished books are not similarly rewarded. Books of numbered cards give a score of one, whereas books of face cards assign a score of two, a book of aces gives a score of three. As a result, when the game ends, the player with the highest score wins.

The game state is determinized by removing from the non-visible player's all card drawn from the deck, shuffling the deck and re-drawing the non-visible player's hand. This means that whenever a card was obtained from the opponent it is no longer treated as invisible, because it cannot be anywhere else than in the opponent's hand or visible on the table in a finished book.

Lost Cities is a two-player card game, designed in 1999 by Reiner Knizia. The goal of the game is to achieve the most profitable set of expeditions to one or more of five lost cities. Players start expeditions by placing numbered cards on them, each player can start up to five expeditions regardless of the opponent's expeditions. Each card in the game has a color and a number, the colors represent one of the five expeditions, the numbers representing the score gained. Next to these cards, colored investment cards cumulatively double the score awarded for an expedition. The deck consists of 60 cards, nine numbered cards per color, and three investment cards per color.

Placing a card on an empty expedition 'initializes' it with a cost of 20. Or, when an investment card is played, with a score of $20 \times I_c$, where I_c is the number of investment cards played on expedition c. These cards can only be played on an expedition when no other cards have been played on it. For example, playing the 'red 5' card starts the red expedition with a cost of 20 and a score of 5 resulting in a -15 score for the player. With a single investment card on this expedition, the score will be 30. Playing more cards on the expedition leads to higher scores. However, only increasing cards may be placed on top of others. In this example, the card 'red 3' can no longer be played, whereas the 'red 8' card can be played.

Each turn, players may either play or discard a card, and draw a card from the draw pile or one of the discard piles. Discarding a card places it on top of one of the colored discard piles which are accessible to both players. The game ends when no cards are left on the draw pile, the player with the highest score wins the game.

In Lost Cities, interaction between players is limited. However, players have to carefully choose their expeditions partly based on their opponent's choices. Moreover, players must be careful not to discard cards which may benefit their opponent, but at the same time take care that they can draw cards beneficial to their chosen expeditions.

As in Go Fish, the game state is determinized by removing the non-visible player's hand, shuffling the deck and re-drawing the non-visible player's hand.

4.2 Phantom Games

Next, we describe two so-called phantom games, Phantom Domineering and Phantom Go. Phantom games are modified versions of fully observable games, in which part of the game state is made invisible to the players. Both games are otherwise fully deterministic, *i.e.*, no roll of the dice, or drawing cards. Consequently, whenever a player makes a move it may be rejected, the player may move again until his move is no longer rejected. Playing a move that is rejected is always beneficial, since it provides the player with new information of the actual game state.

Phantom Domineering is based on the combinatorial game Domineering, which is generally played on a square board with two players. Each turn players block two adjacent positions on the board, one player plays vertically, and the other horizontally. The game ends when one of the players cannot make move. As with most combinatorial games, the first player unable to make a move loses the game, and draws are not possible.

In Phantom Domineering, players can only directly observe their own pieces on the board. For both players, their opponent's pieces are hidden, and can only be observed indirectly by performing rejected moves. A unique property in Phantom Domineering is that rejected moves do not provide immediate information about the opponent's moves. In games where moves consist of occupying single positions, a rejected move can immediately reveal an opponent's move. In Phantom Domineering, however, a rejected move means that either one of the two positions is blocked, or both. Therefore, when determinizing, all opponent's stones are first replaced such that they match the rejected moves, after this, all remaining stones are placed randomly on the board.

Phantom Go is a version of Go played in which the opponent's stones are not revealed. When a move is illegal it is usually because there is an opponent's stone on the chosen intersection. In this case a referee publicly announces that the player made an illegal move and the same player may move again. The Chinese rules are used for scoring games. Phantom Go is played by humans at Go congresses and is enjoyed by spectators who can see both players' boards as well as the complete referee board.

During determinization opponent stones are placed on illegal moves. The remaining opponent stones are placed randomly on the determinized board [9]. The principle of our engine, GoLois, is to perform one play-out per determinization. For each possible move, a large number of determinizations followed by play-outs is performed. The move with the highest average is then chosen. Using this approach, GoLois won the gold medal in 5 of the 6 Phantom Go tournaments held during the last Computer Olympiads.

5 Experiments and Results

In this section we show the results of the experiments performed on four, partially observable two-player games. H-MCTS and the games were implemented in two different engines. Go Fish, Lost Cities and Phantom Domineering are implemented in a Java based engine. Phantom Go is implemented in the $C++$ based engine GoLois.

Lost Cities relies heavily on a heuristic play-out strategy which prevents obvious bad moves such as starting an expedition without a chance of making a profit. These heuristics improve play over a random play-out by up to 40 %.

Table 1. Win rates with respect to the row player. Minimum of 1,000 games per experiment, 10,000 simulations per move.

	Go Fish			
	H-MCTS	SH	MCTS	UCB
H-MCTS	-	**60.9 %** ± 2.9	**54.3 %** ± 1.9	**62.3 %** ± 2.9
SH	39.1 % ± 2.9	-	44.0 % ± 3.0	51.3 % ± 2.0
MCTS	45.7 % ± 1.9	**56.0 %** ± 3.0	-	**55.0 %** ± 3.1
UCB	37.7 % ± 2.9	48.7 % ± 2.0	45.0 % ± 3.1	-
	Lost Cities			
	H-MCTS	SH	MCTS	UCB
H-MCTS	-	46.1 % ± 3.1	**54.1 %** ± 3.1	47.1 % ± 3.1
SH	**53.9 %** ± 3.1	-	**55.6 %** ± 1.9	50.1 % ± 1.9
MCTS	45.9 % ± 3.1	44.4 % ± 1.9	-	45.3 % ± 3.1
UCB	52.9 % ± 3.1	49.9 % ± 1.9	**54.7 %** ± 3.1	-
	8×8 Phantom Domineering			
	H-MCTS	SH	MCTS	UCB
H-MCTS	-	45.1 % ± 3.1	**59.9 %** ± 3.0	**59.5 %** ± 3.0
SH	**54.9 %** ± 3.1	-	**55.1 %** ± 3.1	**58.6 %** ± 3.1
MCTS	41.1 % ± 3.0	44.9 % ± 3.1	-	49.4 % ± 3.1
UCB	40.5 % ± 3.0	41.4 % ± 3.1	51.6 % ± 3.1	-

Table 2. Win rates with respect to the row player. Minimum of 1,000 games per experiment, 25,000 simulations per move.

	Go Fish			
	H-MCTS	SH	MCTS	UCB
H-MCTS	-	**62.2 %** ± 2.9	**55.2 %** ± 3.0	**61.8 %** ± 2.9
SH	42.2 % ± 3.0	-	42.2 % ± 3.0	51.7 % ± 3.1
MCTS	44.9 % ± 3.0	**57.9 %** ± 3.0	-	**59.0 %** ± 3.0
UCB	38.2 % ± 2.9	48.3 % ± 3.1	41.0 % ± 3.1	-
	Lost Cities			
	H-MCTS	SH	MCTS	UCB
H-MCTS	-	48.6 % ± 1.9	**52.7 %** ± 1.9	44.9 % ± 3.0
SH	51.4 % ± 1.9	-	**57.6 %** ± 3.1	52.8 % ± 3.1
MCTS	47.4 % ± 1.9	42.4 % ± 3.0	-	43.7 % ± 1.9
UCB	**55.1 %** ± 3.1	47.3 % ± 3.1	**56.3 %** ± 1.9	-
	8 × 8 Phantom Domineering			
	H-MCTS	SH	MCTS	UCB
H-MCTS	-	48.9 % ± 3.1	53.0 % ± 3.1	**54.5 %** ± 3.1
SH	51.1 % ± 3.1	-	**56.1 %** ± 3.1	51.8 % ± 3.1
MCTS	47.0 % ± 3.1	43.9 % ± 3.1	-	51.3 % ± 3.1
UCB	45.6 % ± 3.1	48.7 % ± 3.1	51.3 % ± 3.1	-

In Phantom Domineering, an ϵ-greedy play-out strategy selects moves based on the number of available moves for the opponent and the player to move. It chooses the move that maximizes their difference. For both Go Fish and Phantom Go, moves are selected uniformly random during play-outs.

In the next subsection, we run experiments on the test domains using a set of different algorithms:

- **H-MCTS** selects moves according to Sequential Halving at the root and UCT in all other parts of the tree, according to Algorithm 2. In all domains, single observer IS-MCTS [13] is used.
- **SH** selects among available moves according to Sequential Halving (Algorithm 1), and samples the moves by play-out immediately. As such, no search is performed.
- **MCTS** selects moves using UCT from root to leaf. As in H-MCTS, single observer IS-MCTS is used.
- **UCB** selects among available moves according to the UCT selection method (Eq. 1) and samples the move immediately by play-out. As such, no search is performed. The method is similar to using the UCB1 algorithm for MABs.

Table 3. Experimental results for Phantom Go. SH vs. UCB with varying C constant. 1,000 games, win rates with respect to SH.

	C	Phantom Go	
		10,000 Simulations	25,000 Simulations
SH vs. UCB	0.1	**69.6 % ± 2.9**	**70.6 % ± 2.9**
	0.2	**58.8 % ± 3.1**	**58.6 % ± 3.1**
	0.3	**58.1 % ± 3.1**	**54.3 % ± 3.1**
	0.4	**58.0 % ± 3.1**	53.0 % ± 3.1
	0.5	**57.1 % ± 3.1**	**55.3 % ± 3.1**
	0.6	**59.1 % ± 3.1**	51.9 % ± 3.1
	0.7	**60.3 % ± 3.1**	**56.7 % ± 3.1**
	0.8	**61.5 % ± 3.1**	**58.6 % ± 3.1**
	0.9	**64.5 % ± 3.0**	**57.8 % ± 3.1**

In all experiments, and for all algorithms, a new determinization is uniformly selected for each simulation. For each individual game, the C constant, used by UCT (Eq. 1) was tuned. MCTS, UCB, and H-MCTS use the same value for the C constant in all experiments.

5.1 Results

For each table, the results are shown with respect to the row algorithm, along with a 95 % confidence interval. For each experiment, the players' seats were swapped such that 50 % of the games are played as the first player, and 50 % as the second, to ensure no first-player or second-player bias. Because H-MCTS cannot be terminated any-time we present only results for a fixed number of simulations. In each experiment, both players are allocated a budget of either 10,000, or 25,000 play-outs. In all tables, significantly positive results are bold-faced.

Tables 1 and 2 show the comparative results for search performed with 10,000 and 25,000 simulations per move, respectively. For most experiments in these tables, 1,000 games were played. However, in some cases where results were close to confidence bounds, 1,500 extra games were played. First, results show that only in Go Fish did performing search improve performance over flat Monte-Carlo sampling, in both Lost Cities and Phantom Domineering performing search did not improve performance. This coincides with previous results for Phantom Go, for which it was determined that search could did not perform better than UCB sampling.

In all games, using H-MCTS improves performance over MCTS when sampling 10,000 simulations per move. In the 25,000 case, MCTS and UCB's performances appear to recover in Lost Cities and Phantom Go. In Go Fish, performance is stable with respect to the number of simulations. For the games where performing search does not improve performance over single-ply sampling, SH is either on par or outperforms UCB.

In all cases, in both experimental setups, SH or H-MCTS either outperforms MCTS and UCB significantly, or does not negatively impact performance. In Phantom Domineering, sampling using SH improves performance over UCB by up to 8.6 %. A significant improvement when considering that no knowledge or heuristics were introduced in the search. Moreover, SH improves the performance of the award-winning engine GoLois by up to 7.1 % over UCB, as shown in Table 3. In this table we detail the results over different C constants for UCB, showing that without tuning any parameter, SH is able to outperform UCB in all cases. UCB's performance similarly somewhat recovers when given more simulations. However, in all but two cases (when $C = 0.4$ or $C = 0.6$), SH still significantly outperforms UCB with 25,000 simulations per move.

6   Conclusions

This paper has investigated Sequential Halving as a selection policy in partially observable games. In the MCTS framework, Sequential Halving was applied at the root of the search tree, and UCB1 elsewhere, leading to a hybrid algorithm called H-MCTS. Experimental results revealed that H-MCTS performed the best in Go Fish, whereas its performance is on par in Lost Cities and Phantom Domineering. In Phantom Go, Sequential Halving as a flat Monte-Carlo Search was the best algorithm for 10,000 play-outs. For 25,000 play-outs, it was still competitive but the difference with the alternative approach UCB was not statistically significant. Even in cases where Sequential Halving was not better it still has the advantage that it is parameter free.

A possible cause for concern when using UCT in partially observable domains is that the statistics in the tree may become conditioned on a set of determinizations. When a new determinization is used for each sample, the current statistics of each node are biased towards previous determinizations and may not necessarily hold for other determinizations in the future. A uniform selection method such as Sequential Halving may circumvent this possible problem, since selection is not based on the current statistics of each node. Rather, nodes are explored uniformly regardless of their statistics and are only removed from selection after being sampled equally often as their siblings.

7   Future Research

Based on the results in this paper and previous work [20], H-MCTS and Sequential Halving have shown promising result in both fully and partially observable games. This leads to several directions for future research. We propose an investigation into SHOT and H-MCTS in partially observable games by using a limited set of determinizations, and a single tree per determinization. Because in these cases it is possible to use Sequential Halving at internal nodes other than the root. For future work in H-MCTS in general, the All-Moves-As-First (AMAF) [5] heuristic is considered, a popular method used in MCTS to improve early estimation of nodes. For partially observable domains in specific, we intend to investigate non-uniform selection of determinizations.

References

1. Arneson, B., Hayward, R., Henderson, P.: Monte-Carlo tree search in Hex. IEEE Trans. Comput. Intell. AI Games **2**(4), 251–258 (2010)
2. Audibert, J., Bubeck, S., Munos, R.: Best arm identification in multi-armed bandits. In: Proceedings of the 23rd Conference on Learning Theory, pp. 41–53 (2010)
3. Auer, P., Cesa-Bianchi, N., Fischer, P.: Finite-time analysis of the multiarmed bandit problem. Mach. Learn. **47**(2–3), 235–256 (2002)
4. Balla, R.K., Fern, A.: UCT for tactical assault planning in real-time strategy games. In: Boutilier, C. (ed.) Proceedings of the 21st International Joint Conference on Artificial Intelligence (IJCAI), pp. 40–45 (2009)
5. Bouzy, B., Helmstetter, B.: Monte-Carlo Go developments. In: van den Herik, H.J., Iida, H., Heinz, E.A. (eds.) Advances in Computer Games. IFIP, vol. 135, pp. 159–174. Springer, New York (2004)
6. Browne, C., Powley, E., Whitehouse, D., Lucas, S.M., Cowling, P.I., Rohlfshagen, P., Tavener, S., Perez, D., Samothrakis, S., Colton, S.: A survey of Monte-Carlo tree search methods. IEEE Trans. Comput. Intell. AI Games **4**(1), 1–43 (2012)
7. Bubeck, S., Munos, R., Stoltz, G.: Pure exploration in finitely-armed and continuous-armed bandits. Theor. Comput. Sci. **412**(19), 1832–1852 (2010)
8. Buro, M., Long, J., Furtak, T., Sturtevant, N.: Improving state evaluation, inference, and search in trick-based card games. In: Boutilier, C. (ed.) Proceedings of the 21st International Joint Conference on Artificial Intelligence, IJCAI 2009, Pasadena, CA, USA, pp. 1407–1413 (2009)
9. Cazenave, T.: A phantom-go program. In: van den Herik, H.J., Hsu, S.-C., Hsu, T., Donkers, H.H.L.M.J. (eds.) CG 2005. LNCS, vol. 4250, pp. 120–125. Springer, Heidelberg (2006)
10. Cazenave, T.: Sequential halving applied to trees. IEEE Trans. Comput. Intell. AI Games **7**(1), 102–105 (2015)
11. Ciancarini, P., Favini, G.: Monte Carlo tree search in Kriegspiel. AI J. **174**(11), 670–6684 (2010)
12. Coulom, R.: Efficient selectivity and backup operators in Monte-Carlo tree search. In: van den Herik, H.J., Ciancarini, P., Donkers, H.H.L.M.J. (eds.) CG 2006. LNCS, vol. 4630, pp. 72–83. Springer, Heidelberg (2007)
13. Cowling, P., Powley, E., Whitehouse, D.: Information set Monte Carlo tree search. IEEE Trans. Comput. Intell. AI Games **4**(2), 120–143 (2012)
14. Feldman, Z., Domshlak, C.: Simple regret optimization in online planning for Markov decision processes. J. Artif. Intell. Res. (JAIR) **51**, 165–205 (2014)
15. Ginsberg, M.: Gib: Steps toward an expert-level bridge-playing program. In: Dean, T. (ed.) Proceedings of the Sixteenth International Joint Conference on Artificial Intelligence (IJCAI 1999), vol. 1, pp. 584–589. Morgan Kaufmann (1999)
16. Karnin, Z., Koren, T., Somekh, O.: Almost optimal exploration in multi-armed bandits. In: Proceedings of the International Conference on Machine Learning, pp. 1238–1246 (2013)

17. Kocsis, L., Szepesvári, C.: Bandit based Monte-Carlo planning. In: Fürnkranz, J., Scheffer, T., Spiliopoulou, M. (eds.) ECML 2006. LNCS (LNAI), vol. 4212, pp. 282–293. Springer, Heidelberg (2006)
18. Nijssen, J.A.M., Winands, M.H.M.: Monte-Carlo tree search for the hide-and-seek game Scotland Yard. Trans. Comput. Intell. AI Games **4**(4), 282–294 (2012)
19. Pepels, T., Winands, M.H.M., Lanctot, M.: Real-time Monte-Carlo tree search in Ms Pac-Man. IEEE Trans. Comp. Intell. AI Games **6**(3), 245–257 (2014)
20. Pepels, T., Cazenave, T., Winands, M.H.M., Lanctot, M.: Minimizing simple and cumulative regret in Monte-Carlo tree search. In: Cazenave, T., Winands, M.H.M., Björnsson, Y. (eds.) CGW 2014. CCIS, vol. 504, pp. 1–15. Springer, Heidelberg (2014)
21. Powley, E.J., Whitehouse, D., Cowling, P.I.: Monte Carlo tree search with macro-actions and heuristic route planning for the physical travelling salesman problem. In: IEEE Conference on Computational Intelligence and Games, pp. 234–241. IEEE (2012)
22. Rimmel, A., Teytaud, O., Lee, C., Yen, S., Wang, M., Tsai, S.: Current frontiers in computer Go. IEEE Trans. Comput. Intell. AI Games **2**(4), 229–238 (2010)
23. Russell, S., Norvig, P.: Artificial Intelligence: A Modern Approach, 3rd edn. Prentice-Hall Inc., Upper Saddle River (2010)
24. Sheppard, B.: World-championship-caliber Scrabble. Artif. Intell. **134**(1–2), 241–275 (2002)
25. Tolpin, D., Shimony, S.: MCTS based on simple regret. In: Proceedings of the Association for the Advancement Artificial Intelligence, pp. 570–576 (2012)
26. Winands, M.H.M., Björnsson, Y., Saito, J.T.: Monte Carlo tree search in lines of action. IEEE Trans. Comp. Intell. AI Games **2**(4), 239–250 (2010)

An Experimental Investigation
on the Pancake Problem

Bruno Bouzy[✉]

LIPADE, Université Paris Descartes, Paris, France
bruno.bouzy@parisdescartes.fr

Abstract. In this paper, we present an experimental investigation on the pancake problem. Also called sorting by prefix reversals (SBPR), this problem is linked to the genome rearrangement problem also called sorting by reversals (SBR). The pancake problem is a NP-hard problem. Until now, the best theoretical R-approximation was 2 with an algorithm, which gives a 1.22 experimental R-approximation on stacks with a size inferior to 70. In the current work, we used a Monte-Carlo Search (MCS) approach with nested levels and specific domain-dependent simulations. First, in order to sort large stacks of pancakes, we show that MCS is a relevant alternative to Iterative Deepening Depth First Search (IDDFS). Secondly, at a given level and with a given number of polynomial-time domain-dependent simulations, MCS is a polynomial-time algorithm as well. We observed that MCS at level 3 gives a 1.04 experimental R-approximation, which is a breakthrough. At level 1, MCS solves stacks of size 512 with an experimental R-approximation value of 1.20.

1 Introduction

The pancake problem is described as follows. A chef prepares a stack of pancakes that come out all different sizes on a plate. The goal of the server is to order them with decreasing sizes, the largest pancake touching the plate, and the smallest pancake being at the top. The server can insert a spatula below a pancake and flip the substack situated above the spatula. He may repeat this action as many times as necessary. In the particular version, the goal of the server is to sort a particular stack with a minimum number of flips. In the global version, the question is to determine the maximum number of flips $f(n)$ - the diameter - to sort any stack of n pancakes.

This problem is a puzzle, or a one-player game well-known in artificial intelligence and in computer science under the name of sorting by prefix reversals (SBPR). Its importance is caused by its similarity with the sorting by reversals (SBR) problem which is fundamental in biology to understand the proximity between genomes of two different species. For example, the SBR distance between a cabbage and a turnip is three [23]. The SBR problem has been studied in depth [24] for the last twenty years. The SBR problem can be signed when the signs of the genes are considered, or unsigned otherwise. Similarly, the pancakes can be burnt on one side, or not. This brings about four domains: one with

© Springer International Publishing Switzerland 2016
T. Cazenave et al. (Eds.): CGW 2015/GIGA 2015, CCIS 614, pp. 30–43, 2016.
DOI: 10.1007/978-3-319-39402-2_3

unburnt pancakes, one with burnt pancakes, one with unsigned genes and one with signed genes.

In the unburnt pancake problem, Gates and Papadimitriou [22] gave the first bounds of the diameter in 1979, and Bulteau has shown that the problem is NP-hard in 2013 [11]. Very interesting work have been done between 1979 and today. The goal of the current work is to provide an experimental contribution to the unburnt pancake problem. More specifically, we show the pros and cons of two planning algorithms used in computer games: IDDFS [30] and MCS [13]. Besides, we define several domain-dependent algorithms: Efficient Sort (EffSort), Alternate Sort (AltSort), Basic Random EFficient algorithm (BREF), Fixed-Depth Efficient Sort (FDEffSort), and Fixed-Depth Alternate Sort (FDAltSort), and we re-use the Fischer and Ginzinger's algorithm (FG) [21]. FG was proved to be a 2-approximation algorithm that also reaches a 1.22 approximation experimentally. We show how to use the algorithms above in the MCS framework. We obtain an experimental approximation of 1.04, which is a significant reduction.

The paper is organized as follows. Section 2 defines the SBR and SBPR problems. Section 3 sums up the related work in the four domains. Section 4 presents our work and its experimental results. Section 5 concludes.

2   Definitions

Let N be the size of a permutation π and

$$[\pi(1), \pi(2), ..., \pi(N-1), \pi(N)]$$

the representation of π. The problem of sorting a permutation by reversals consists in reaching the identity permutation

$$[1, 2, ..., N-1, N]$$

by applying a sequence of reversals. A reversal $\rho(i, j)$ with $i < j$ is an action applied on a permutation. It transforms the permutation

$$[\pi(1), ..., \pi(i-1), \pi(i), ..., \pi(j), \pi(j+1)..., \pi(N)]$$

into

$$[\pi(1), ..., \pi(i-1), \underline{\pi(j), ..., \pi(i)}, \pi(j+1)..., \pi(N)].$$

The effect of a reversal is reversing the order of the numbers between the two cuts. A cut is located between two numbers of the permutation. In the above example, the first cut is between $i-1$ and i and the second one is between j and $j+1$.

In the pancake problem, each number $\pi(i)$ corresponds to the size of the pancake situated at position i in a stack of pancakes, and the permutation problem is seen as a pancake stack to be sorted by decreasing size. One cut is fixed and corresponds to the top of the stack. The other cut corresponds to the location of a spatula inserted between two pancakes so as to reverse the substack above the spatula. For example, the permutation

$$[\pi(1), ..., \pi(i), \pi(i+1)..., \pi(N)]$$

has its top on the left and its bottom on the right. After a flip $\rho(i)$ between i and $i + 1$, the permutation becomes

$$[\underline{\pi(i), ..., \pi(1)}, \pi(i + 1)..., \pi(N)].$$

In addition, a permutation can be signed or not. In the signed case, a sign is associated to each number, i.e. the integers of the permutation can be positive or negative. When performing a reversal, the sign of the changing numbers changes too. For example, after the reversal $\rho(i, j)$,

$$[\pi(1), ..., \pi(i - 1), \pi(i), ..., \pi(j), \pi(j + 1)..., \pi(N)]$$

becomes

$$[\pi(1), ..., \pi(i - 1), \underline{-\pi(j), ..., -\pi(i)}, \pi(j + 1)..., \pi(N)].$$

The burnt pancake problem is the signed version of the pancake problem. In the burnt pancake problem, the pancakes are burnt on one side, and a flip performs the reversal and changes the burnt side. The goal is to reach the sorted stack with all pancakes having their burnt side down.

In the literature, the permutations are often extended with two numbers, $N + 1$ after $\pi(N)$, and 0 before $\pi(1)$, and the extended representation of permutation π is

$$[0, \pi(1), ..., \pi(N), N + 1].$$

The reversal distance of a permutation π is the length of the shortest sequence of reversals that sorts the permutation.

A basic and central concept in SBR problems is the breakpoint. For $1 \leq i \leq N + 1$, a breakpoint is situated between i and $i - 1$ when $|\pi(i) - \pi(i - 1)| \neq 1$. In the following, $\#bp$ is the number of breakpoints. Since each breakpoint must be removed to obtain the identity permutation, and since one reversal removes at most one breakpoint, $\#bp$ is a lower bound of the reversal distance. In the planning context, $\#bp$ is a simple and admissible heuristic. In the pancake problem, the possible breakpoint between the top pancake and above is not taken into account. In the signed permutation problem or in the burnt pancake problem a breakpoint is situated between i and $i - 1$ when $\pi(i) - \pi(i - 1) \neq 1$.

3 Related Work

Since the four domains are closely linked, this section presents them in order: signed permutation, unsigned permutation, unburnt pancakes, and burnt pancakes.

3.1 Signed Permutations

The best overview of the genome rearrangement problem to begin with is by Hayes [24]. Since the genes are signed, the genome rearrangement problem is mainly connected with the signed permutation problem, but also to the unsigned permutation problem. In 1993, Bafna and Pevzner [4] introduced the cycle graph. In 1995, Hannenhalli and Pevzner [23] devised the first polynomial-time algorithm for signed permutations. Its complexity was in $O(n^4)$. The authors introduced the breakpoint graph, and the so called hurdles. This work is the reference. The follow up consists in several refinements.

In 1996, Berman and Hannenhalli [6], and Kaplan et al. [27] in 1998, enhanced the result with an algorithm whose complexity was in $O(n^2)$. The concept of fortress was new. In 2001, Bader and colleagues [3] found out an algorithm that finds the reversal distance in $O(n)$, but without giving the optimal reversal sequence. In 2002, GRIMM [39], a web site, was developed to implement the above theories. In 2003, [28] described efficient data structures to cope with the problem. Then, in 2005, Anne Bergeron [5] introduced a simple and self-contained theory, which does not use the complexities of the previous algorithms, and that solves signed permutation problems in quadratic time as well. In 2006, [37,38] are subquadratic improvements.

3.2 Unsigned Permutations

The basic work in the unsigned permutation problem is [29] in 1992, by Kececioglu and Sanker. This problem was proved to be NP-hard [12] by Caprara in 1997. In 1998, Christie [16] described the 3/2-approximation algorithm. Reversal corresponding to red nodes are relevant. Furthermore, the Christie's thesis [17] described many approaches for other classes of permutation problems. In 1999 and 2001, [7,8] contain complexity results. In 2003, [2,36] describe evolutionary approaches to the unsigned permutation problem. Particularly, the work of Auyeung and Abraham [2], performed in 2003, consists in finding out the best signature of an unsigned problem, with a genetic algorithm. The best signature is the signature such that the signed permutation reversal distance is minimal. Computing this distance is performed in linear time [3].

3.3 Unburnt Pancakes

The unburnt pancake problem is the most difficult of the four domains [11]. Related work focused on the diameter of the pancake graph. In 2004, a pancake challenge was set up. Tomas Rokicki won the challenge and gave explanations to solve and generate difficult pancake problems [34]. Fischer and Ginzinger published their 2-approximation algorithm [21].

Bounds on the Diameter. A focus is to bound the diameter of the graph of the problem in N the size of the pancake stack. The first bounds on the pancake problem diameter were found by Gates and Papadimitriou in 1979 [22]: $(5n + 5)/3$ is the upper bound and $(17/16)n$ is the lower bound. To prove the upper bound, [22] exhibits an algorithm with several cases. They count the number of actions corresponding to each case and obtain inequalities. They formulate a linear program whose solution proves the upper bound. To prove the lower bound, they exhibit a length-8 elementary permutation that can be used to build length-n permutations with solutions of length $(18/16)n$ on average but bounded by below by $(17/16)n$. The Gates and Papadimitriou's sequence is

$$GP = [1, 7, 5, 3, 6, 4, 2, 8].$$

The $(15/14)n$ lower bound was found by Heydari and Sudborough in 1997 [26].

In 2006, the diameter of the 17-pancake graph [1] was computed. In 2009, a new upper bound was found on the diameter: $(18/11)n$ [15]. In 2010, in the planning context, the breakpoint heuristic #bp was explicitly used in a depth-first-search [25]. In 2011, [18] Josef Cibulka showed that $17n/12$ flips were necessary to sort stacks of unburnt pancakes on average over the stacks of size n. Josef Cibulka mentioned a list of interesting concepts: deepness, surfaceness, biggest-well-placed, second-biggest-well-placed, smallest-not-on-top.

The 2004 Pancake Challenge. In 2004, a pancake challenge was organized to focus on the resolution of specific problems. In a first stage, the entrants had to submit pancake problems. In the second stage, the entrants had to solve the submitted problems. The entry of the winner of the pancake challenge was the one of Tomas Rokicki [34]. Its entry is described and gives really interesting ideas.

The Inverse Problem and Backward Solutions. Considering π^{-1} the inverse permutation of the original problem π can be helpful. For example, if

$$\pi = [5, 3, 6, 1, 4, 2]$$

then

$$\pi^{-1} = [4, 6, 2, 5, 1, 3].$$

π^{-1} and π correspond to two different problems. The one is the forward problem yielding a forward solution, and the other is the backward problem yielding a backward solution (BS). Because $\pi\pi^{-1} = Id$, the backward solution is the reverse sequence of the forward solution. The enhancement consists in solving the two problems simultaneously and comparing the lengths of the two solutions, and comparing the times to get them. We call it the BS enhancement. For IDDFS, the two solutions are optimal and share the same length, but the times to solve them can be very different. For MCS or approximate algorithms, the lengths of the two sequences can be different, and the idea consists in keeping the shortest solution. BS works in practice. See Table 3 compared to Table 2.

Difficult Positions. In the diameter estimation context, Tomas Rokicki exhibited two elementary sequences:

$$S5 = [1, 3, 5, 2, 4]$$

and
$$L9 = [1, 5, 8, 3, 6, 9, 4, 7, 2].$$
Then, he built L9-based (resp. S5-based) permutations by repeating the L9 (resp. S5) permutation shifted by 9 (resp. by 5). For example,
$$L9(2) = [1, 5, 8, 3, 6, 9, 4, 7, 2, 10, 14, 17, 12, 15, 18, 13, 16, 11].$$
These sequences $L9(x)$ and $S5(y)$ were used to attempt to prove a 11/10 ratio lower bound on the diameter. Unfortunately, this approach did not work. However, these ad hoc sequences are hard to solve, and we consider them as hard problems in the following.

The 2-Approximation Algorithm of Fischer and Ginzinger. Fischer and Ginzinger designed FG, an algorithm that is a 2-approximation polynomial algorithm [21]. It means that the length L_{FG} of the solution found by FG is inferior to two times the length of the optimal solution. Since the best lower bound known today is the number of breakpoints #bp, it means that L_{FG} is proved to be inferior to $2 \times $ #bp. In practice, Fischer and Ginzinger mention an approximation ratio $Rapprox = 1.22$. The idea is to classify moves in four types. Type 1 moves are the ones that remove a breakpoint in one move. Type 2 and type 3 moves lead a pancake to the top of the stack so as to move it to a correct place at the next move. Type 4 moves correspond to the other cases. Fischer and Ginzinger proves that type 2, 3, 4 moves removes a breakpoint in less than 2 moves, and that type 1 moves remove a breakpoint in one move.

Pancake Flipping Is Hard. In 2012, Laurent Bulteau and his colleagues proved that the pancake flipping problem is NP-hard [10,11]. He did this by exhibiting a polynomial algorithm that transforms a pancake problem into a SAT problem and vice-versa. He gave an important clue to solve the pancake problem. He considered sequence of type 1 moves only, i.e. moves removing one breakpoint. He defined efficiently sorted permutations, i.e. permutations that can be sorted by type 1 moves, or "efficient" moves. He defined deadlock permutations without type 1 move. A sequence of type 1 moves reaches either the identity permutation and the permutation is efficiently sorted, or deadlock permutations only, and the permutation is not efficiently sorted. To see whether a permutation is efficiently sortable, a binary tree must be developed. Bulteau made a polynomial correspondence between the efficiently sortable permutation problem and the SAT problem, proving by this translation that the former problem is NP-hard.

Miscellaneous. [33] contains results about the genus of pancake network.

3.4   Burnt Pancakes

Here again, the focus was to bound the diameter too. In 1995, the first bounds on the diameter and a conjecture [19] were presented: $3n/2$ is a lower bound and $2(n-1)$ a upper bound. The second bounds on the diameter were proved in 1997 [26]. A polynomial-time algorithm on "simple" burnt pancake problems [31] was published in 2011. In 2011, Josef Cibulka showed that $7n/4$ flips were necessary to sort stacks of burnt pancakes on average over the stacks of size n [18]. He also disproved the conjecture by Cohen and Blum [19]. Josef Cibulka mentioned interesting concepts: anti-adjacency and clan.

4 Our Work

First, this section presents the domain-independent algorithms used in our work. Secondly, it presents the domain dependent algorithms designed in the purpose of our work. Thirdly, it presents the settings of the experiments. Then, this section yields the results of the experiments in order.

4.1 Domain Independent Algorithms

This section describes the algorithms we used to solve pancake problems as efficiently as possible. There are two basic and general algorithms:

- Iterative Deepening Depth-First Search (IDDFS) [30],
- Monte-Carlo Search (MCS) [13].

We consider IDDFS as an exponential-time algorithm in N [30]. When it completes, the solution found is optimal. However, for N superior to a threshold, IDDFS needs to much time, and becomes useless actually. However, before completion, IDDFS yields a lower bound on the optimal length.

MCS [13] is a simulation-based algorithm that gave very good results in various domains such as general game playing [32], expression discovery [14], morpion solitaire [35], weak Schur numbers [20] and cooperative path-finding [9]. MCS is used with a level L. At any time, MCS stores its best sequence found so far, thus it yields an upper bound on the optimal solution. When used at a given level L, MCS is a polynomial-time algorithm in N. Let assume that the level 0 simulations are polynomial-time algorithms. Let T_0 be the time used to perform a complete level 0 simulations. Let us bound T_1 the time to complete a level 1 MCS simulation. To move one step ahead in a level 1 simulation, MCS launches at most N level 0 simulations, which costs $N \times T_0$. Since the length of level 0 simulation is bounded by $2N$, we have $T_1 \leq 2N^2T_0$. If T_0 is polynomial in N then T_1 is polynomial as well. By induction, a level L MCS is polynomial-time. The higher L, the higher the polynomial degree.

The threshold effect observed for IDDFS does not appear for a polynomial-time algorithm. If you obtain solutions for N in time T, and if d is the degree of the polynomial, you may obtain solutions for $N+1$ in time $T \times (N+1)^d/N^d$ which is just a little bit more expensive than T. Therefore, we get two tools to work with: one is costly but optimal when it completes, IDDFS, and the other one is approximate but its cost is polynomial-time, MCS.

4.2   Domain Dependent Algorithms

We have designed several pancake problem dependent algorithms: Efficient Sort (EffSort), Alternate Sort (AltSort), Basic Random Efficient algorithm (BREF), Fischer and Ginzinger algorithm (FG) [21], Fixed-Depth Efficient Sort (FDEffSort), Fixed-Depth Alternate Sort (FDAltSort). We have implemented each of them and we describe them briefly here.

Since a position has at most two efficient moves, EffSort searches within a binary tree to determine whether a permutation is efficiently sortable [11] or not. If the permutation is efficiently sortable, the sequence of efficient moves is output, and the output permutation is the identity. Otherwise, the longest sequence of efficient moves is output, and the output permutation is a deadlock (i.e. a permutation without efficient move).

When a position is a deadlock, the solver has to perform an inefficient move, i.e. a move that does not lower $\#bp$. A waste move is a move that keeps the $\#bp$ constant. We define two kinds or waste moves: hard or soft. A waste move is hard if it creates a breakpoint while removing another one. A waste move is soft otherwise (the set of breakpoints is unchanged). We define a destroying move as a move that increases $\#bp$.

We designed AltSort. While the output permutation is not sorted, AltSort iteratively calls EffSort and performs a soft waste move if the output permutation is a deadlock. When EffSort is successful, AltSort stops. AltSort and EffSort are inferior to IDDFS. They are exponential time algorithms.

We designed BREF that iteratively chooses and performs an efficient move if possible. A position may have 0, 1 or 2 efficient moves. On a position with one efficient move, this move is selected with probability 1. On a position with two efficient moves, one of them is selected with probability 0.5. Otherwise, BREF chooses and performs a soft waste move defined above. Most of the times, a position without efficient moves has a lot of soft waste moves available. In this case, the move is chosen at random with uniform probability. At the end of a simulation, the reward is the length of the simulation. BREF is a randomized version of AltSort. BREF and FG are polynomial-time algorithms. They can serve as level 0 simulation for MCS.

FDEffSort is the fixed-depth version of EffSort. With depth D, FDEffSort becomes a polynomial-time algorithm. FDAltSort is the version of AltSort using FDEffSort. FDAltSort is a polynomial-time algorithm. It can be used as a level 0 simulation for MCS.

4.3   Experimental Settings

The experiments show the effect of using:

- IDDFS or MCS,
- FG and BREF within MCS,
- backward solutions BS in addition to original solutions,
- FDAltSort within MCS.

There are different classes of test positions: positions randomly generated for a given size, and difficult positions mentioned by related work, mainly [22,26,34]. IDDFS may easily solve easy positions of size 60 randomly generated in a few seconds. However, IDDFS cannot solve some hard positions of size 30. Whatever the size and the problem difficulty, MCS always finds a preliminary solution quickly. This solution is refined as time goes on to become near-optimal or optimal.

We mention three indicators to evaluate our algorithms.

- Finding the minimal length $Lmin$ of an optimal solution for a given problem.
- Finding $Rapprox$ as low as possible with a polynomial time algorithm averaged over a set of 100 problems.
- Limiting the running time of an experiment with 100 problems to one or two hours.

$Rapprox$ is the ratio of the length L of the actual solution found over $Lmin$. Since $Lmin$ is unknown in practice, $Lmin$ is replaced by $\#bp$. $Rapprox = L/\#bp$. The standard deviation of the $Rapprox$ value that we observed for one problem generated at random is roughly 0.05. The two-sigma rule says that the standard deviation over 100 problems is $0.05 \times 2/10 = 0.01$. The values of $Rapprox$ given below are 0.01 correct with probability 0.95. On average, the time to solve one problem is inferior to one minute.

Table 1. IDDFS: In practice, how $Rapprox$ varies in N. L is the average length of solutions. T is the average time in seconds to sort one stack.

N	L	$Rapprox$	T
8	6.5	1.09	0
16	14.5	1.05	0
32	31.0	1.03	0
64	63.0	1.02	5

4.4 MCS and IDDFS

The first experiment consists in assessing IDDFS and MCS under reasonable time constraints: at most one hour. MCS uses BREF as level 0 simulations. When using IDDFS, Table 1 shows how $Rapprox$ varies in N. First, although IDDFS gives an optimal result, $Rapprox$ is not 1. This happens because $\#bp$ is not the length of optimal solutions but a lower bound only. Secondly, Table 1 shows that IDDFS cannot give results in reasonable time for $N > 64$. For $N = 30$, IDDFS does not solve some difficult positions in less than few hours. Thirdly, Table 2 shows how $Rapprox$ varies in N and in level with MCS. Level 0 simulations can be launched easily with $N = 256$. $Rapprox$ is 1.30 for level 0 simulations, 1.28 for level 1 simulations. Then, as the level increases, $Rapprox$ decreases. Level 2 simulations yields $Rapprox = 1.22$ which equals the value mentioned in [21].

Table 2. MCS+BREF: How *Rapprox* varies in N and Level. $L(x)$ is the average length of solutions at level x. $T(x)$ is the average time in seconds to sort one stack at level x.

N	$L(0)$	$R(0)$	$T(0)$	$L(1)$	$R(1)$	$T(1)$	$L(2)$	$R(2)$	$T(2)$	$L(3)$	$R(3)$	$T(3)$	$L(4)$	$R(4)$	$T(4)$
8	7.5	1.15	0	7.5	1.19	0	7.0	1.15	0	6.7	1.12	0	6.5	1.08	0.01
16	18	1.30	0	18	1.27	0	16.5	1.18	0	16.0	1.14	0.01	15.5	1.09	0.05
32	38	1.30	0	37.5	1.26	0	37	1.22	0.02	36.5	1.21	0.6			
64	82	1.32	0	79.5	1.28	0.01	76	1.24	1.2						
128	165	1.30	0.01	163	1.29	0.23									
256	339	1.34	0.05	333	1.31	9									

Table 3. How *Rapprox* varies in N and MCS level with the trick of Backward Solutions (BS). $L(x)$ is the average length of solutions at level x. $T(x)$ is the average time in seconds to sort one stack at level x.

N	$L(0)$	$R(0)$	$T(0)$	$L(1)$	$R(1)$	$T(1)$	$L(2)$	$R(2)$	$T(2)$	$L(3)$	$R(3)$	$T(3)$	$L(4)$	$R(4)$	$T(4)$
8	7.0	1.15	0	6.8	1.14	0	6.7	1.12	0	6.7	1.11	0	6.6	1.10	0.01
16	16.5	1.18	0	16.4	1.18	0	15.6	1.12	0	15.4	1.10	0.01	15.3	1.09	0.1
32	37.0	1.22	0	36.4	1.21	0	35.4	1.17	0.05	35.2	1.15	0.8			
64	79.5	1.28	0.01	77	1.24	0.02	75.2	1.20	2						
128	163	1.29	0.02	159	1.26	0.5									
256	335	1.32	0.1												

For higher levels, *Rapprox* = 1.10 showing that high levels of MCS give good results in reasonable time, and that MCS is a viable alternative to IDDFS even for difficult positions. When compared to *Rapprox* = 1.22 of [21], *Rapprox* = 1.08 is a first breakthrough.

4.5   MCS + BREF + BS

Table 3 shows how *Rapprox* varies in N and in the MCS level when level 0 simulations are the best of one forward simulation and one backward simulation. We call this the BS enhancement. Table 3 must be compared to Table 2. One can observe that the BS enhancement is effective at level 0 indeed, and also at level 1 and level 2. However, its effect is less visible at higher levels of MCS: *Rapprox* = 1.09.

4.6   MCS + FG + BS

Table 4 shows how *Rapprox* varies in N and in the MCS level when level 0 simulations are the forward and the backward FG simulations. So as to see the effect of using FG instead of BREF in MCS, Table 4 must be compared to Table 3. First, the comparison shows that FG is worse than BREF for level 0. FG is on a par with BREF for level 1. For level 2 and higher levels, FG surpasses BREF: *Rapprox* = 1.05.

Table 5 shows how *Rapprox* varies in N and in the MCS level when level 0 simulations are the forward and the backward *randomized* FG. Randomized FG works as follows. If type 1 moves exist, one of them is chosen at random.

Table 4. How *Rapprox* varies in N and MCS level with simulations being FG with the trick of Backward Solutions (BS). $L(x)$ is the average length of solutions at level x. $T(x)$ is the average time in seconds to sort one stack at level x.

N	L(0)	R(0)	T(0)	L(1)	R(1)	T(1)	L(2)	R(2)	T(2)	L(3)	R(3)	T(3)	L(4)	R(4)	T(4)
8	7.73	1.28	0	6.85	1.14	0	6.7	1.10	0	6.7	1.10	0.01	6.7	1.10	0.01
16	18.2	1.30	0	16.0	1.14	0	15.1	1.07	0.01	14.9	1.05	0.06	14.8	1.05	0.6
32	39.7	1.29	0	35.2	1.16	0.01	32.6	1.08	0.12	31.6	1.05	4.2			
64	82.4	1.32	0.02	74.5	1.20	0.06	68.5	1.10	2.5						
128	167	1.33	0.04	155	1.23	0.8									
256	336	1.32	0.2												

Table 5. How *Rapprox* varies in N and MCS level with simulations being randomized FG with the trick of Backward Solutions (BS). $L(x)$ is the average length of solutions at level x. $T(x)$ is the average time in seconds to sort one stack at level x.

N	L(0)	R(0)	T(0)	L(1)	R(1)	T(1)	L(2)	R(2)	T(2)	L(3)	R(3)	T(3)	L(4)	R(4)	T(4)
8	8.4	1.40	0	7.0	1.16	0	6.7	1.10	0	6.7	1.10	0.01	6.7	1.10	0.01
16	19.6	1.40	0	16.1	1.14	0	15.2	1.08	0.01	14.9	1.05	0.08	14.8	1.06	0.6
32	43.8	1.45	0	35.1	1.16	0.01	32.4	1.07	0.2	31.5	1.04	4			
64	90	1.43	0.02	76.5	1.23	0.1	67.5	1.08	6						
128	181	1.43	0.04	160	1.26	1.3									
256	356	1.40	0.3												

Otherwise, if type 2 moves exist, one of them is chosen at random and so on. So as to see the effect of using randomized FG instead of FG, Table 5 must be compared to Table 4. As expected and as shown by the $R(0)$ column of Tables 4 and 5, randomized FG yields a worse *Rapprox* than direct FG. Randomized FG is worse than direct FG for level 1. However, when used in higher levels of MCS, *Rapprox* with the randomized version is slightly inferior to *Rapprox* with the direct version. At level 4, MCS gives *Rapprox* = 1.05 when using randomized version of FG as basic simulations.

4.7   MCS + FDAltSort

In a preliminary experiment, not reported here, we assessed MCS using AltSort directly as level 0 simulations. This did not work well on hard positions because AltSort is not a polynomial-time algorithm. Consequently some simulations did not complete quickly. We had to limit the depth at which EffSort searches and we had to create FDAltSort. (FDEffSort determines whether a permutation is efficiently sortable at depth D). FDAltSort can be used as a level 0 simulation in MCS. We assessed MCS using FDAltSort at depth $D = 10$. For each level, Table 6 displays the variations of *Rapprox* in N. These results must be compared to the results of Table 5. At level 0, FDAltSort is better than randomized FG and on a par with FG. At level 1, level 2 and level 3, FDAltSort is better than the other algorithms. Launching FDAltSort at level 4 was not interesting. However, *Rapprox* achieves 1.04 at its minimal value. Furthermore, the good point here is that the results are obtained for pancake stack sizes going up to 512 instead

Table 6. How *Rapprox* varies in N and MCS level with simulations being FDAltSort at depth $D = 10$. $L(x)$ is the average length of solutions at level x. $T(x)$ is the average time in seconds to sort one stack at level x.

N	$L(0)$	$R(0)$	$T(0)$	$L(1)$	$R(1)$	$T(1)$	$L(2)$	$R(2)$	$T(2)$	$L(3)$	$R(3)$	$T(3)$
8	6.7	1.12	0	6.6	1.09	0	6.6	1.09	0	6.6	1.09	0
16	16.0	1.15	0	14.8	1.08	0	14.5	1.04	0.01	14.5	1.04	0.04
32	37.3	1.23	0	33.3	1.10	0.02	31.8	1.05	0.1	31.3	1.04	0.8
64	80.4	1.29	0	69.7	1.12	0.05	66.4	1.06	1	65.1	1.04	18
128	164	1.30	0.01	145	1.15	0.3	139	1.10	10.5			
256	326	1.28	0.02	300	1.18	1.5						
512	647	1.27	0.06	614	1.20	9						

of 256 before, and with *Rapprox* = 1.20. This is a significant improvement. FDAltSort as simulations are much more efficient than FG or BREF were. We also tried to incorporate the BS enhancement, but the results were worse.

5   Conclusion and Future Work

In this work, we summed up the state-of-the-art of the permutation sorting by reversals domain. This domain was studied in depth by researches on the genome. It remains fascinating as underlined by Hayes [24]. The unsigned permutation domain is hard [12] but the signed permutation domain has polynomial-time solver [23]. The pancake problem was less studied. The unburnt pancake problem is difficult [11] while the complexity of the burnt pancake is unknown.

Our contribution is experimental. It shows how MCS extends the results obtained by IDDFS on unburnt pancake stacks. On the one hand, IDDFS can solve some pancake stacks of size 60 [25] in a few minutes but cannot solve some specific hard pancake stacks [34] of size 30 only. On the other hand, MCS can solve pancake stacks of significantly higher sizes and the hard pancake stacks of size 30 not solved by IDDFS. Practically, our MCS solver solves pancake stacks of size up to 512 with a 1.20 R-approximation, under the best configuration. MCS may use BREF, FG or FDAltSort with results that are approximately equal in terms of running time and *Rapprox* value. Practically, we observed that MCS approximates the best solutions with a *Rapprox* ratio of 1.04 for size up to 64, which is significantly better than the 1.22 value of [21]. Our MCS solver solves pancake stacks of size 128 or 256 with a R-approximation value roughly situated between 1.10 and 1.25.

In a near future, we want to study the burnt pancake problem. Furthermore, the burnt pancake problem is linked to the unburnt pancake problem. A block of sorted unburnt pancakes can be replaced by one burnt pancake, and the unburnt pancake problem becomes a mixed pancake problem. Solutions to burnt pancake problems could be used to solve ending unburnt pancake problems. We want to investigate in this direction.

To date, the number of breakpoints remains the simplest and the most efficient heuristic to bound the optimal solution length by below. However, this admissible heuristic should be refined to better approximate the optimal solution length. Some hard problems - or stacks - are hard because they contain substacks whose solution lengths are strictly higher than the number of breakpoints. As the permutation problems contain concepts such as hurdles or fortresses [6], and as shown by the work of Josef Cibulka on burnt pancakes, we have to find out the corresponding concepts to design appropriate heuristic functions for the pancake problems.

References

1. Asai, S., Kounoike, Y., Shinano, Y., Kaneko, K.: Computing the diameter of 17-pancake graph using a PC cluster. In: Nagel, W.E., Walter, W.V., Lehner, W. (eds.) Euro-Par 2006. LNCS, vol. 4128, pp. 1114–1124. Springer, Heidelberg (2006)
2. Auyeung, A., Abraham, A.: Estimating genome reversal distance by genetic algorithm. In: The 2003 Congress on Evolutionary Computation (CEC 2003), vol. 2, pp. 1157–1161. IEEE (2003)
3. Bader, D., Moret, B., Yan, M.: A linear-time algorithm for computing inversion distance between signed permutation with an experimental study. In: WADS, pp. 365–376 (2001)
4. Bafna, V., Pevzner, P.: Genome rearrangements and sorting by reversals. In: FoCS (1993)
5. Bergeron, A.: A very elementary presentation of the Hannenhalli-Pevzner theory. DAM **146**(2), 134–145 (2005)
6. Berman, P., Hannenhalli, S.: Fast sorting by reversal. In: 7th Symposium on Combinatorial Pattern Matching, pp. 168–185 (1996)
7. Berman, P., Hannenhalli, S., Karpinski, M.: 1.375-approximation algorithm for sorting by reversals. Technical report, 41, DIMACS (2001)
8. Berman, P., Karpinski, M.: On some tighter inapproximability results. Technical report, 23, DIMACS (1999)
9. Bouzy, B.: Monte-Carlo fork search for cooperative path-finding. In: Cazenave, T., Winands, M.H.M., Iida, H. (eds.) Workshop on Computer Games (CGW 2013), vol. 408, pp. 1–15. CCIS (2013)
10. Bulteau, L.: Algorithmic aspects of genome rearrangements. Ph.D. thesis, Université de Nantes (2013)
11. Bulteau, L., Fertin, G., Rusu, I.: Pancake flipping is hard. In: Rovan, B., Sassone, V., Widmayer, P. (eds.) MFCS 2012. LNCS, vol. 7464, pp. 247–258. Springer, Heidelberg (2012)
12. Caprara, A.: Sorting by reversals is difficult. In: ICCMB, pp. 75–83 (1997)
13. Cazenave, T.: Nested Monte-Carlo search. In: IJCAI, pp. 456–461 (2009)
14. Cazenave, T.: Nested Monte-Carlo expression discovery. In: ECAI, pp. 1057–1058. Lisbon (2010)
15. Chitturi, B., Fahle, W., Meng, Z., Morales, L., Shields, C.O., Sudborough, I.H., Voit, W.: A (18/11)n upper bound for sorting by reversals. TCS **410**, 3372–3390 (2009)
16. Christie, D.: A 3/2 approximation algorithm for sorting by reversals. In: 9th SIAM Symposium on Discrete Algorithms (1998)

17. Christie, D.: Genome rearrangement problems. Ph.D. thesis, University of Glasgow (1998)
18. Cibulka, J.: Average number of flips in pancake sorting. TCS **412**, 822–834 (2011)
19. Cohen, D., Blum, M.: On the problem of sorting burnt pancakes. DAM **61**(2), 105–120 (1995)
20. Eliahou, S., Fonlupt, C., Fromentin, J., Marion-Poty, V., Robilliard, D., Teytaud, F.: Investigating Monte-Carlo methods on the weak Schur problem. In: Middendorf, M., Blum, C. (eds.) EvoCOP 2013. LNCS, vol. 7832, pp. 191–201. Springer, Heidelberg (2013)
21. Fischer, J., Ginzinger, S.W.: A 2-approximation algorithm for sorting by prefix reversals. In: Brodal, G.S., Leonardi, S. (eds.) ESA 2005. LNCS, vol. 3669, pp. 415–425. Springer, Heidelberg (2005)
22. Gates, W., Papadimitriou, C.: Bounds for sorting by prefix reversal. Discrete Math. **27**, 47–57 (1979)
23. Hannenhalli, S., Pevzner, P.: Transforming cabbage into turnip: polynomial algorithm for sorting signed permutations by reversals. J. ACM **46**(1), 1–27 (1995)
24. Hayes, B.: Sorting out the genome. Am. Sci. **95**, 386–391 (2007)
25. Helmert, M.: Landmark heuristics for the pancake problem. In: SoCS, pp. 109–110 (2010)
26. Heydari, M., Sudborough, H.: On the diameter of the pancake problem. J. Algorithms **25**, 67–94 (1997)
27. Kaplan, H., Shamir, R., Tarjan, R.E.: Faster and simpler algorithm for sorting signed permutationd by reversals. In: Proceedings of the Eighth Annual ACM-SIAM Symposium on Discrete Algorithms (SODA 1997), pp. 344–351 (1997)
28. Kaplan, H., Verbin, E.: Efficient data structures and a new randomized approach for sorting signed permutations by reversals. In: Baeza-Yates, R., Chávez, E., Crochemore, M. (eds.) CPM 2003. LNCS, vol. 2676, pp. 170–185. Springer, Heidelberg (2003)
29. Kececioglu, J., Sankoff, D.: Exact and approximation algorithms for sorting by reversals with application to genome rearrangement. Algorithmica **13**, 180–210 (1992)
30. Korf, R.: Depth-first iterative-deepening: an optimal admissible tree search. Artif. Intell. **27**(1), 97–109 (1985)
31. Labarre, A., Cibulka, J.: Polynomial-time sortable stacks of burnt pancakes. TCS **412**, 695–702 (2011)
32. Méhat, J., Cazenave, T.: Combining UCT and nested Monte-Carlo search for single-player general game playing. IEEE Trans. Comput. Intell. AI Games **2**(4), 271–277 (2010)
33. Nguyen, Q., Bettayeb, S.: On the genus of pancake nerwork. IAJIT **8**(3), 289 (2011)
34. Rokicki, T.: Pancake entry (2004). http://tomas.rokicki.com/pancake/
35. Rosin, C.D.: Nested rollout policy adaptation for Monte Carlo- tree search. In: IJCAI, pp. 649–654 (2011)
36. Soncco-Alvarez, J.L., Ayala-Rincon, M.: A genetic approach with a simple fitness function for sorting unsigned permutations by reversals. In: 7th Colombian Computing Congress (CCC). IEEE (2012)
37. Tannier, E., Bergeron, A., Sagot, M.F.: Advances on sorting by reversals. DAM **155**(6–7), 881–888 (2006)
38. Tannier, E., Sagot, M.F.: Sorting by reversals in subquadratic time. In: SCPM (2004)
39. Tesler, G.: GRIMM: genome rearrangements web server. Bioinformatics **18**(3), 492–493 (2002)

485 – A New Upper Bound for Morpion Solitaire

Henryk Michalewski$^{(\boxtimes)}$, Andrzej Nagórko, and Jakub Pawlewicz

Faculty of Mathematics, Informatics, and Mechanics,
University of Warsaw, Warsaw, Poland
{H.Michalewski,A.Nagorko,J.Pawlewicz}@mimuw.edu.pl

Abstract. In previous research an upper bound of 705 was proved on
the number of moves in the 5T variant of the Morpion Solitaire game.
We show a new upper bound of 485 moves. This is achieved in the following way: we encode Morpion 5T rules as a linear program and solve
126912 instances of this program on special octagonal boards. In order
to show correctness of this method we analyze rules of the game and use
a concept of a potential of a given position. By solving continuous-valued
relaxations of linear programs on these boards, we obtain an upper bound
of 586 moves. Further analysis of original, not relaxed, mixed-integer programs leads to an improvement of this bound to 485 moves. However,
this is achieved at a significantly higher computational cost.

1 Introduction

The Morpion Solitaire is a paper-and-pencil single-player game played on a
square grid with the initial configuration of 36 dots depicted in Fig. 1. In the
5T variant of the Morpion Solitaire game[1], in each move the player puts a dot
on an unused grid position and draws a line that consists of four consecutive
segments passing through the dot. The line must be horizontal, vertical or diagonal. None of the four segments used in the line may appear as a segment of any
other line. The goal is to find the longest possible sequence of moves.

The problem is notoriously difficult for computers. For 34 years, in the
Morpion 5T game the longest known sequence was one of 170 moves discovered
by C.-H. Bruneau in 1976 (see[2] [1]). The record was finally broken by Christo-
pher D. Rosin, who presented in [6] a configuration of 177 moves, obtained using
a Monte Carlo algorithm called Nested Rollout Policy Adaptation (NRPA). In
2011 (see [1]) he improved his record to 178, which is the best result known
today. The webpage [1] maintained by Christian Boyer, contains an extensive
and up-to-date information about records in all Morpion Solitaire variants.

As the Morpion 5T game is played on a potentially infinite grid, a priori
it is not clear whether the maximal sequence should be finite at all. An upper
bound of 705 was shown in [2]. In the present paper we show a new upper bound

---

[1] We refer to this variant as the Morpion 5T game. For an overview of other variants
see the webpage [1] or the paper [2].

[2] Regarding this and other records we refer to the webpage [1] for a detailed description
and further references.

© Springer International Publishing Switzerland 2016
T. Cazenave et al. (Eds.): CGW 2015/GIGA 2015, CCIS 614, pp. 44–59, 2016.
DOI: 10.1007/978-3-319-39402-2_4

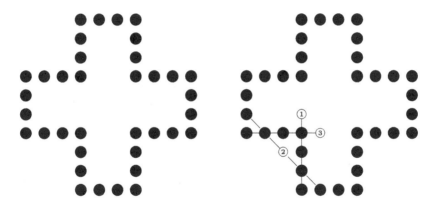

Fig. 1. Initial position of Morption 5T on the left and a position with 3 moves on the right.

of 485. We base our approach on the observation that a Morpion 5T game may be expressed as a mixed-integer linear programming problem.

We observed (see Sect. 2.1) that the bound obtained by a continuous-valued linear relaxation of the Morpion 5T game significantly depends on the size of the grid on which the game is played. On big grids the bound may be well over the upper bound of 705 (in fact, we do not know if it is finite). On the other hand on smaller grids we obtained useful bounds. For example, on a grid limited to a regular octagon[3] with sides of length 10, we obtain a linear relaxation bound of 543. By a geometric reasoning, using the notion of potential and careful analysis of the rules of Morpion 5T game, we show that every Morpion 5T position must be contained in one of 126912 octagonal grids with small boundaries (see Lemma 4 for a precise formula). The maximum bound of 586.82 is obtained on a grid which is an octagon with sides of lengths $10, 8, 10, 12, 10, 8, 10, 12$ (see Fig. 5).

In fact, the picture is more complicated. To state a linear problem for Morpion 5T we need not only to consider the shape of the octagonal grid but also a position of the initial cross inside. This makes the number of cases to consider larger by two orders of magnitude. To get around this difficulty we consider variants of Morpion Solitaire called Morpion 5T+ and Morpion 5T++ (see [1]). In the later variant the position of the initial cross inside of the grid is not relevant. Every Morpion 5T game is also a Morpion 5T+ game and every Morpion 5T+ game is a Morpion 5T++ game. The difference between 5T and 5T+ is that the line drawn in a move needs not to pass through the dot placed in this move. The difference between 5T+ and 5T++ is that one may place more than one dot in a single move, as long as in the final position the number of dots is equal to the number of moves plus 36. That is, we may borrow dots in 5T++ as long as the balance at the end is correct. In Morpion 5T++ we start with an empty board.

The upper bound of 705 moves proved in [2] is also valid in the Morpion 5T++ game. The longest known sequence of moves in Morpion 5T+ was found

---

[3] This is a graph that plays a role in the proof of the 705 bound in [2].

by Marc Bertin in 1974 and consists of 216 moves (see [1]). The longest known sequence of moves in the Morpion 5T++ game was found by Christian Boyer in 2011 (see [1]). The sequence consists of 317 moves.

The associated mixed-integer linear problems are much easier to solve in the case of 5T++ variant and we have a benefit of much smaller number of cases to consider. However, the limit on the size of the grid pertains to the 5T variant, so our new upper bound of 485 is valid for the 5T variant only.

The paper is organized as follows. In Sect. 2 we formulate the linear problem (LP0)–(LP3). In Sect. 3 we calculate that the number of instances, which must be treated by the solver, is 126912. In Sect. 4 we consider consequences of the relaxation of the original problem (LP0)–(LP3). This allows to show an upper bound of 586 moves in the Morpion 5T game. In Sect. 5 we push the result of Sect. 4 in order to obtain an upper bound of 485 moves. This is done at a considerable increase in the computation time. In Sect. 6 we explain the correctness of algorithms used in previous Sections. The correctness result boils down to an observation how, in terms of potential, a given board relates to the smallest octagon containing this board (see Theorem 1).

We note that modern LP solvers have no problem in finding the record sequence of 317 moves for the 5T++ variant, but despite considerable computational effort we were not able to break this record. The current upper bound of 485 can be improved with more computational resources. However, we believe that the best approach would be to find better limitations on the size of grids.

It is also possible to write linear programs that solve the Morpion 5D variant of the Morpion Solitaire (see [1] for description of the rules and current lower and upper bounds). On larger grids we obtain objective 144 for the relaxed problem, as the standard potential-based argument applies to the relaxed case as well (see[4] [2]). The upper bound of 121 moves in the Morpion 5D game was proved in [4]. Using variants of the Morpion 5D game and a different strategy of limiting grids, we were able to prove that an upper bound in the Morpion 5D game is below 100. We also proved that the best possible result in the symmetric Morpion 5D game is 68. These results will be presented in a separate publication.

2   Linear Relaxation

A *lattice point* on a plane is a point with integer coordinates. A *lattice graph* is a graph with vertices in lattice points and edges consisting of pairs (p, q), where p and q are two different neighboring points, that is $p \neq q$ and $p = (n, m)$ and $q = (n \pm i, m \pm j)$ for some $i, j = 0, 1$. We call such edges the *lattice edges*.

A *move* in a lattice graph $G = (V, E)$ is a set of four consecutive parallel lattice edges. We let $\mathcal{M}(G)$ to be the set of all moves in a graph G. We start with the following observation, which simply rephrases the rules of Morpion 5T++ formulated in the Introduction.

Lemma 1. *A graph $G = (V, E)$ is a Morpion 5T++ position graph if and only if it satisfies the following conditions*

---
[4] In fact, applying additional argumentation, in [2] is shown a bound of 141 moves.

(M1) G is a lattice graph,
(M2) $4 \cdot \#V - \#E = 144$,
(M3) The set E of edges of G can be decomposed into a collection of disjoint moves.

Let $B = (V_B, E_B)$ be a fixed lattice graph that we shall call *the board*. In applications, it will be a sufficiently large octagonal lattice graph with a full set of edges. Below we define linear constraints that describe all subgraphs of B that satisfy conditions (M1)–(M3) of Lemma 1.

We define the following set of structural *binary* variables, that is variables assuming values $0, 1$:

$$\{\text{dot}_v : v \in V_B\} \cup \{\text{mv}_m : m \in \mathcal{M}(B)\}. \tag{LP1}$$

For each $e \in E_B$ and $v \in e$ we declare the following constraints:

$$\sum_{m \in \mathcal{M}(B), e \in m} \text{mv}_m \leq \text{dot}_v. \tag{LP2}$$

$$\sum_{v \in V_B} \text{dot}_v = 36 + \sum_{m \in \mathcal{M}(B)} \text{mv}_m. \tag{LP3}$$

The following two lemmas describe correspondence between binary-valued solutions of a mixed integer programming problem (LP1)–(LP3) and subgraphs of B that are Morpion 5T++ positions.

Lemma 2. *Let $G = (V_G, E_G)$ be a subgraph of B and a Morpion 5T++ position obtained by a sequence $\mathcal{M}$ of moves. If*

$$\text{dot}_v = \begin{cases} 0 & \text{if } v \notin V_G \\ 1 & \text{if } v \in V_G \end{cases} \quad \text{and} \quad \text{mv}_m = \begin{cases} 0 & \text{if } v \notin \mathcal{M} \\ 1 & \text{if } v \in \mathcal{M} \end{cases},$$

then conditions (LP1), (LP2) and (LP3) hold.

Proof. If $\text{dot}_v = 0$, then there is no move passing through v, hence the left hand side of (LP2) is equal to 0. If $\text{dot}_v = 1$, then condition (LP2) means that every segment e played in the game can appear in exactly one move. Condition (LP3) means that the number of dots placed is higher by 36 than the number of moves made.

Lemma 3. *Assume that a set of variables defined by condition (LP1) satisfies conditions (LP2) and (LP3). Let $G = (V_G, E_G)$ be a graph with a set of vertices*

$$V_G = \{v \in V_B : \text{dot}_v = 1\}$$

and a set of edges

$$E_G = \{e \in E_B : \exists_{m \in \mathcal{M}(B)} \ e \in m, \text{mv}_m = 1\}.$$

Then G is a Morpion 5T++ position and a subgraph of G.

Proof. We will show that G satisfies conditions (M1)–(M3) of Lemma 1.

By the definition of E_G, if $e \in E_G$ then there exists $m \in \mathcal{M}(B)$ such that $\mathrm{mv}_m = 1$. By (LP2), if $\mathrm{mv}_m = 1$, then $\mathrm{dot}_v = 1$ for each $v \in V_B$ such that $v \in e \in m$. It means that graph G contains vertices of its edges, therefore it is a well defined subgraph of B, hence it is a lattice graph and it satisfies (M1).

From (LP2) follows, that the moves mv_m must be disjoint in the sense, that they cannot contain the same edge twice. This implies condition (M3) of Lemma 1. From disjointness and condition (LP3) follows condition (M2) of Lemma 1.

We consider a linear relaxation of the MIP problem (LP1)–(LP3). We let structural variables to be real-valued, subject to bounds

$$0 \leq \mathrm{dot}_v, \mathrm{mv}_m \leq 1. \tag{LP4}$$

In the relaxation we maximize the objective function

$$\sum_{m \in \mathcal{M}(B)} \mathrm{mv}_m \tag{LP0}$$

Clearly, an optimal solution to the linear programming problem (LP0)–(LP4) gives an upper bound for the length of a Morpion 5T++ game on a board B.

2.1   The Problem of the Infinite Grid

Observe that any lattice graph that consists of 9 vertex-disjoint moves has 45 vertices and 36 edges and satisfies conditions (M1)–(M3) of Lemma 1, hence it is a Morpion 5T++ position graph and consequently Morpion 5T++ positions can have arbitrarily large diameter in the plane $\mathbb{R}^2$.

The following table summarizes solutions of the linear relaxation of Morpion 5T++ on square $n \times n$ boards (where n is the number of edges on the side) (Fig. 2).

10	20	30	40	50	60	70	80	90	100
64.00	278.50	619.53	876.55	1130.01	1387.54	1641.74	1898.13	2152.86	2408.54

Fig. 2. The top row contains the length n of the edge of a given square and the bottom row contains solutions to the relaxed problem (LP0)–(LP4) on the $n \times n$ board.

We do not know whether the objective function (LP0) is bounded or not on the infinite grid. However, the bound of 705 moves derived in [2] holds for Morpion 5T++. This shows that we get no useful upper bound for positions satisfying (M1)–(M3) using our linear relaxation method. To get a bound, we have to use another properties of Morpion 5T positions to bound the size of the board. This will be done in the next Section.

3 Bounding the Board

Let $G = (V, E)$ be a lattice graph. Following [2] we define the *potential of G*

$$\text{potential}(G) = 8 \cdot \#V - 2 \cdot \#E.$$

In this Section it will appear that the missing constraint which makes the original linear problem (LP0)–(LP4) accessible to modern LP solvers is an additional bound on the shape and potential of the board (see Theorem 1), which in turn, thanks to Lemma 4, will imply a bound on the size of relevant boards. In order to formulate the bound we need some new geometric notions.

A *half–plane graph* is a full lattice graph with a vertex set of all lattice points $\langle x, y \rangle$ such that

$$ax + by + c \geq 0$$

where $a, b \in \{-1, 0, 1\}$ with $a \neq 0$ or $b \neq 0$ and $c \in \mathbb{Z}$.

An *octagonal hull* of a lattice graph $G = (V_G, E_G)$, denoted $\text{hull}(G)$, is an intersection of all half–plane graphs containing G. We call octagonal hulls *octagons*.

Every octagon has eight edges. We may describe octagon by giving lengths of its edges. We start with the top edge and continue clockwise. For example, octagon depicted in Fig. 3 has edges of lengths 3, 3, 0, 1, 6, 0, 3, 1. We call every diagonal edge of length 0 a corner of an octagon. Corners are opposite if they correspond to parallel edges of the octagon.

In the next two Sections we will obtain an upper bound of 586 and respectively 485 moves for Morpion 5T game solving 126912 instances of the linear problem (LP0)–(LP4) described in Sect. 2. The following Theorem shows that the penalty, measured in extra potential, paid for solving such problems only on octagonal hulls is relatively small. In turn, thanks to Lemma 4, the bound on the potential allows to limit the size of octagons. Thus we can focus attention on 126912 relatively small octagonal instances of linear programs. The number 126912 will be deduced in Theorem 2 later in this Section.

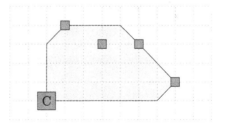

Fig. 3. The octagonal hull of green points. The green point marked with the letter C is a *corner* of the octagon hull. (Color figure online)

Theorem 1. *Let G be a Morpion 5T position graph.*

1. If $\text{hull}\,G$ *does not contain any corner, then*

$$\text{potential}(\text{hull}(G)) \leq \text{potential}(G)$$

2. If it does not contain opposite corners, then

$$\text{potential}(\text{hull}(G)) \leq \text{potential}(G) + 2$$

3. If it contains at least two opposite corners, then

$$\text{potential}(\text{hull}(G)) \leq \text{potential}(G) + 4.$$

We define $\text{modifier}(\text{hull}(G))$ *as* 0 *in case 1,* 2 *in case 2 and* 4 *in case 3.*

We postpone the proof of Theorem 1 until Sect. 6.

3.1   The Set of Boards

In this Subsection we describe a set of octagonal boards that contain every possible Morpion 5T position. As this set is quite large, we will use symmetry to limit its size.

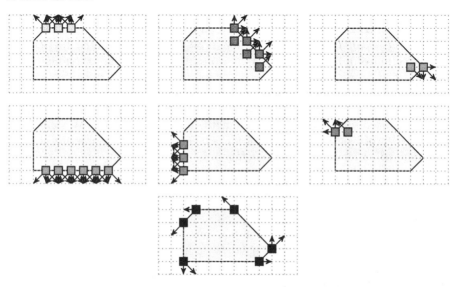

Fig. 4. The octagon $(3, 3, 0, 1, 6, 0, 3, 1)$. The top 6 figures represent potential associated with $3a_1 = 3 \cdot 3 = 9$, $4a_2 = 4 \cdot 3 = 12$, $4a_4 = 4 \cdot 1 = 4$, $3a_5 = 3 \cdot 6$, $3a_7 = 3 \cdot 3 = 9$, $4a_8 = 4 \cdot 1 = 4$. The bottom figure represents missing 8 edges of the potential.

We say that a graph G is *non-degenerated* if it contains three vertices that are not on a single diagonal line.

Lemma 4. *Let G be a non-degenerated octagon with edges of length $a_1, a_2, a_3, a_4, a_5, a_6, a_7, a_8$ with a_1 denoting the length of the top edge. The following equations hold.*

$$\text{potential}(G) = 8 + 3a_1 + 4a_2 + 3a_3 + 4a_4 + 3a_5 + 4a_6 + 3a_7 + 4a_8. \quad \text{(O1)}$$

$$a_8 = a_2 + a_3 + a_4 - a_7 - a_6. \quad \text{(O2)}$$

$$a_1 = a_4 + a_5 + a_6 - a_8 - a_2. \quad \text{(O3)}$$

If G contains the starting cross of Morpion 5T game, then

$$a_1 + a_2 + a_3 \geq 10 \qquad (O4)$$

$$a_8 + a_1 + a_2 \geq 10 \qquad (O5)$$

$$a_2 + a_3 + a_4 \geq 10 \qquad (O6)$$

Using rotation by multiple of 90 degrees and a reflection along the y-axis we can always obtain a graph that satisfies

$$a_1 \geq a_3, a_1 \geq a_5, a_1 \geq a_7 \qquad (O7)$$

$$a_8 \geq a_2 \qquad (O8)$$

Proof. Condition (O1) follows from distribution of potential visualized in Fig. 4. Conditions (O2), (O3) are elementary geometric properties of octagons. We will verify them on the example presented in Fig. 4. Indeed,

$$a_8 = 1.$$

On the other hand

$$a_2 + a_3 + a_4 - a_7 - a_6 = 3 + 0 + 1 - 3 - 1 = 1.$$

Similarly,

$$a_1 = 3$$

and

$$a_4 + a_5 + a_6 - a_8 - a_2 = 1 + 6 + 0 - 1 - 3 = 3.$$

Properties (O4), (O5) and (O6) follows from the observation that in order to embed the starting cross of Morpion 5T (see Fig. 1), the projections of the octagon in diagonal, horizontal and vertical directions must be of length at least 10.

Property (O7) can be guaranteed through rotation by a multiple of 90 degrees. Then property (O8) can be guaranteed through reflection along the y–axis. This reflection preserves property (O7).

Theorem 2. *Let $\mathcal{O}$ denote the set of octagons O that satisfy constraints (O1)–(O8) of Lemma 4 and the constraint given by the equality* potential$(O) = 288 +$ modifier(O). *The number of elements of $\mathcal{O}$ is 126912 and the octagon with the largest number of vertices is an equilateral octagon with sides of length 10. This octagon contains 741 vertices.*

Proof. Every octagon O with potential$(O) < 288 +$ modifier(O) is included in an octagon O' with potential$(O') = 288 +$ modifier(O'). Hence we can ignore in our calculations octagons with potential$(O) < 288 +$ modifier(O). The number of relevant octagons was calculated using the script `octagons.cpp` (see the repository [5]). The script generates all instances of octagons satisfying constraints of this Theorem.

As a corollary we obtain the bound presented in [2].

Corollary 1 ([2]) *The number of moves in a Morpion 5T game is bounded by 705.*

Proof. We list all octagons in $\mathcal{O}$ and check how many dots can placed in a fixed octagon, given the starting 36 dots. The best result consists of 705 new dots for the equilateral octagon with sides of length 10.[5]

Let us notice, that this Corollary is weaker than the one obtained in [2], because the above method applies to Morpion 5T, but not to Morpion 5T++ game.

4   Linear Relaxation Bound

Theorem 3. *Let* obj_B *denote the value of optimal solution of a linear programming problem (LP0)–(LP3). If B satisfies conditions (O1)–(O8) of Lemma 4, then*

$$\mathrm{obj}_B \leq 586.82353.$$

Fig. 5. The record illustrating maximal solution of the relaxed problem. This solution is obtained for the octagon $(10, 8, 10, 12, 10, 8, 10, 12)$. Since this is a relaxed problem (LP0)–(LP4), the grayness in the Figure indicates the value of the move, that is a number between 0 and 1.

---

[5] This proof does not rely on linear optimization. We just go over a finite list of octagons.

The maximum value is obtained for the octagon

$$B = (10, 8, 10, 12, 10, 8, 10, 12).$$

The record solution can be found in Fig. 5. All 126912 relaxed problems were solved by **gurobi** optimization software (see [3]) within 24 hours on a single core of a Linux machine equipped with **Intel**® **Xeon**® CPU X3220@2.40GHz with 8GB of RAM.

Proof. The proof is a calculation of obj_B for all octagons satisfying conditions (O1)–(O8) of Lemma 4. The source code of the program **generator.cpp** generating the relevant linear programs can be found in the repository [5].

Corollary 2. *The number of moves in a Morpion 5T game is bounded by 586.*

5   MIP Bound: 485

Let us notice that results obtained in Theorem 3 can be naturally strengthened through longer computations. In practice, we were able to target the objective 485.

From Sect. 4 we know all 126912 instances and their performance under relaxed (LP0)–(LP4) linear problem. Apparently, out of all 126912 problems, 42889 instances have the relaxed bound bigger or equal to 485. These are exactly the instances which must be treated by direct computations if we want to reduce the bound to 485. The total computation time for

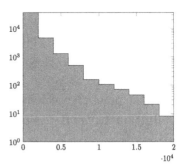

Fig. 6. The vertical axis shows the number of cases and the horizontal axis shows the computation time in seconds.

this target was approximately 310 days using the optimization software **gurobi** (see [3]) on a single core of a Linux machine equipped with **Intel**® **Xeon**® CPU X3220@2.40GHz with 8GB of RAM. The graph 6 shows on the logarithmic scale the distribution of the computation time among 42889 instances[6].

6   Geometry of the Problem

In this Section we will show a proof of Theorem 1. The key technical ingredient is Lemma 10. Let $\mathcal{M} = \mathcal{M}(G)$ be the set of all possible moves in a graph G. A lattice graph with a vertex set V is *full* if its edge set is maximal, i.e.

$$E = \{(u, v) \colon u, v \in V, u \neq v, |u_x - v_x| \leq 1, |u_y - v_y| \leq 1\}.$$

A graph $G = (V, E)$ is *1-connected* if a full lattice graph with a vertex set V is connected. A *boundary* of a lattice graph $G = (V, E)$ is a set

$$\mathcal{B}(G) = \{\langle u, v \rangle \in \mathbb{Z}^2 \times \mathbb{Z}^2 \colon u \in V, (u, v) \notin E\}.$$

Observe that elements of $\mathcal{B}(G)$ are directed edges with start points in V such that the corresponding undirected edge is not in E. Let us notice, that

---

[6] In fact, a half of the instances required less than 100 s to reach the limit of 485 and 9 instances required the computation time longer than 18000 s.

$$\text{potential}(G) = \#\mathcal{B}(G).$$

where $\text{potential}(G)$ is the number defined at the beginning of Sect. 3. Here we will analyze the potential more closely and divide it into *external* and *internal* potentials.

An edge $e = \langle u, v \rangle \in \mathcal{B}(G)$ is an *external edge* if $u + k \cdot (v - u) \notin V$ for every $k \geq 1$. Let $\mathcal{B}^{\text{ex}}(G)$ denote the set of all external edges of G.

An edge $e \in \mathcal{B}(G)$ is an *internal edge* if it is not an external edge. Let $\mathcal{B}^{\text{int}}(G)$ denote the set of all internal edges of G.

The *external potential* $\text{potential}^{\text{ex}}(G)$ is the cardinality of the set $\mathcal{B}^{\text{ex}}(G)$. The *internal potential* $\text{potential}^{\text{int}}(G)$ is the cardinality of the set $\mathcal{B}^{\text{int}}(G)$.

Lemma 5. *If G is a Morpion position graph, then* $\text{potential}(G) = 288$.

Proof. From the definition of potential at the beginning of Sect. 3 we have

$$\text{potential}(G) = 8\#V - 2\#E.$$

The number 288 for Morpion position graphs follows from property (M2) in Lemma 1.

In the proof of Theorem 1 we need some additional definitions.

Definition 1. *Let $\mathcal{L}$ denote the*

$$\mathcal{L} = \{l_{a,b,c} \colon (a, b) \in \mathcal{D}, c \in \mathbb{Z}\}.$$

where

$$l_{a,b,c} = \{(x, y) \in \mathbb{Z}^2 \colon ax + by + c = 0\}$$

A line $l_{a,b,c} \in \mathcal{L}$ is called diagonal if $\langle a, b \rangle \in \{-1, 1\}^2$. A graph $G = (V, E)$ is degenerated if there exists a line $l \in \mathcal{L}$ such that $V \subset l$.

Definition 2. *A line $l \in \mathcal{L}$ is a gap line for graph G if l is diagonal, does not contain any vertices of G, but there are vertices of G on both sides of l. We let $\text{gap}(G)$ be the number of gap lines of a graph G.*

Lemma 6. *If a lattice graph $G = \langle V, E \rangle$ is an octagon, then for every $l \in \mathcal{L}$ the intersection $V \cap l$ is 1-connected and* $\text{potential}(G) = \text{potential}^{\text{ex}}(G)$.

Proof. For every $l_{a,b,c} \in \mathcal{L}$ the intersection in $\mathbb{R}^2$ of $\{(x, y) \in \mathbb{R}^2 : ax + by + c = 0\}$ with the convex hull of G in $\mathbb{R}^2$ is an interval in $\mathbb{R}^2$ and the lattice points of this interval are 1-connected and coincide with $V \cap l$.

Lemma 7. *If G is not degenerated, then* $\text{gap}(\text{hull}(G)) = 0$.

Proof. Fix a diagonal line l. We will show that l is not a gap line for $\text{hull}(G)$. Take vertices $u, v \in \text{hull}(G)$ on both sides of l. Since G is non-degenerated, we may assume that u and v are not on a same line perpendicular to l (first we select arbitrary two points on both sides of l and if they are located on the same line then from degeneracy we can find another either on the side of u or on the side of v).

From the definition, $\text{hull}(G)$ contains the intersection of all half-planes that contain u and v, hence it contains a parallelogram with the following characteristics:

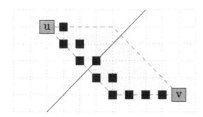

Fig. 7. Every line parallel to l with u and v on different sides must contain one of the vertices marked with black squares.

– opposite vertices of the parallelogram are u and v,

– a pair of edges of parallelogram is horizontal (or vertical)

– another pair of edges is diagonal (see Fig. 7).

We start from the vertex v and mark the black dots along the horizontal/vertical edge. Then along the diagonal edge we mark "the staircase" as in Fig. 7. Notice, that l passes through one of the black dots of the staircase.

Indeed, since u and v are not on a line perpendicular to l, if l intersect the parallelogram between two points on the diagonal edge, then l necessarily pass through a black dot between these two points.

If l intersects one of the horizontal/vertical edges of the parallelogram, then l passes through one of the black dots located on the edge. Hence l has a non–empty intersection with $\mathrm{hull}(G)$ and it follows that l is not a gap line for $\mathrm{hull}(G)$.

The following Lemma provides a sufficient condition for a line in $\mathcal{L}$ to be a gap line.

Lemma 8. *Let G be a 1-connected, bounded and non-degenerated lattice graph and $l \in \mathcal{L}$. If l does not contain a vertex of G and contains a vertex of $\mathrm{hull}(G)$, then l is a gap line for G.*

Proof. Every half-plane graph that contains l must contain at least one vertex of G, since l has a vertex in $\mathrm{hull}(G)$ (otherwise the opposite half-plane with l removed would contain whole G and hence also $\mathrm{hull}(G)$, so $\mathrm{hull}(G)$ would be disjoint from l). Therefore both half-planes that have l as a boundary contain vertices of G. Since G is disjoint with l, there are vertices of G on both sides of l. In order to prove that l is a gap line it is enough to verify that l is diagonal.

Indeed, observe that horizontal and vertical lines disconnect the grid $\mathbb{Z}^2$ into two 1-connected components. Since G contains vertices on both sides of l and is 1-connected, the line l must be diagonal, hence it is a gap line for G.

The above Lemma 8 along with Lemma 7 shows a characterization of gap lines among lines in $\mathcal{L}$.

Lemma 9. *If G is a 1-connected, non-degenerated lattice graph, then*

$$\mathrm{potential}(\mathrm{hull}(G)) = \mathrm{potential}^{\mathrm{ex}}(G) + 2\,\mathrm{gap}(G).$$

Proof. In this proof it will be convenient to mark as $\bar{e}$ the set consisting of two vertices at the ends of a given edge e. Observe that for any graph Γ and any line $l \in \mathcal{L}$

$$\#\{\bar{e}\colon e \in \mathcal{B}^{\mathrm{ex}}(\Gamma),\ \bar{e} \subset l\} \tag{$*$}$$

is 0 iff $V_\Gamma \cap l = \emptyset$ and 2 otherwise. By Lemma 6,

$$\text{potential}(\text{hull}(G)) = \text{potential}^{\text{ex}}(\text{hull}(G)).$$

We have

$$\text{potential}^{\text{ex}}(\text{hull}(G)) = \sum_{l \in \mathcal{L}} \#(l \cap \mathcal{B}^{\text{ex}}(\text{hull}(G)))$$

and

$$\text{potential}^{\text{ex}}(G) = \sum_{l \in \mathcal{L}} \#(l \cap \mathcal{B}^{\text{ex}}(G)).$$

By (*) and by Lemma 8, for a given $l \in \mathcal{L}$ either $\#(l \cap \mathcal{B}^{\text{ex}}(\text{hull}(G))) = \#(l \cap \mathcal{B}^{\text{ex}}(G))$ or l is a gap line for G and $\#(l \cap \mathcal{B}^{\text{ex}}(\text{hull}(G))) = 2$, $\#(l \cap \mathcal{B}^{\text{ex}}(G)) = 0$. Hence

$$\text{potential}(\text{hull}(G)) = \text{potential}^{\text{ex}}(G) + 2\,\text{gap}(G).$$

The main technical difficulty in the Section is the following geometric Lemma. This Lemma together with Lemma 9 finish the proof of Theorem 1.

Lemma 10. *If G is a position of the Morpion 5T then*

$$2\,\text{gap}(G) \le \text{potential}^{\text{int}}(G) + \text{modifier}(\text{hull}(G)).$$

The notion of modifier was defined in Theorem 1.

Proof. Let $G = (V, E)$. Let $\mathcal{L}(G)$ denote the set of all gap lines of G. Let $l \in \mathcal{L}(G)$. The two halfplanes bounded by l decompose the set V of vertices of G into two disjoint subsets, one of which contains all dots of the initial cross. If the other set contains only a single vertex, then we say that l is a singular gap line. Otherwise we say that l is a non-singular gap line. If l is a singular gap line, then we let v_l denote the single vertex of V separated from the initial cross by line l and call v_l the singular vertex of l.

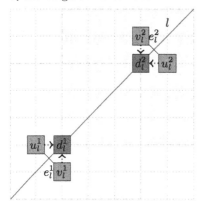

Fig. 8. Green vertices belong to the graph, red vertices do not belong to the graph. From the dotted arrows we will choose 2 to compensate for the gap line l. (Color figure online)

Let $m_1, m_2, \ldots, m_n$ be a sequence of the Morpion 5T moves that lead to a position G. Let l be a singular gap line and let m_k be a move that puts the singular vertex v_l on board. Since there are no other vertices in the halfplane bounded by l that contains v_l, no move in sequence $m_1, m_2, \ldots$ requires dot v_l and the sequence $m_1, m_2, \ldots, m_{k-1}, m_{k+1}, \ldots, m_n$ is a valid move sequence. Likewise $m_1, m_2, \ldots,$ $m_{k-1}, m_{k+1}, \ldots, m_n, m_k$ is valid. Hence we may modify our sequence so that the moves that put singular dots on board are at the very end of the move sequence.

Let $m_1, m_2, \ldots, m_k, m_{k+1}, \ldots, m_n$ be a sequence of Morpion 5T moves that lead to a position G such that moves $m_1, m_2, \ldots, m_k$ put non-singular dots on board and moves

$m_{k+1}, \ldots, m_n$ put singular dots on board. Let $H = (V_H, E_H)$ be a Morpion 5T position obtained by the sequence $m_1, m_2, \ldots, m_k$. We will show that

$$2 \operatorname{gap}(H) \le \operatorname{potential}^{\mathrm{int}}(H).$$

Observe that H has no singular gap lines as removing a singular vertex cannot make a non-singular vertex singular.

Let $l \in \mathcal{L}(H)$. Since l is non-singular and H is obtained as a position in Morpion 5T game, there are two edges $e_l^1, e_l^2 \in E_H$ that cross l. Consider labeling of dots as in Fig. 8. Note that d_l^1 and d_l^2 are picked on l between e_l^1 and e_l^2 (and they may be the same point when e_l^1 and e_l^2 are next to each other).

We will construct a map that assigns to each e_l^i ($i = 1, 2, l \in \mathcal{L}(H)$) one edge from the list

$$(u_l^i, d_l^i), (v_l^i, d_l^i) \qquad\qquad (**_l^i)$$

in such a way that the following conditions are satisfied.

1. The assigned edges realize the internal potential of H, i.e. they belong to $\mathcal{B}^{\mathrm{int}}(H)$.
2. We do not assign the same edge twice.

First we'll show that at least one edge from the edge list $(**_l^i)$ belongs to $\mathcal{B}^{\mathrm{int}}(H)$. Without a loss of generality we may assume that $i = 1$. Consider two half-lines starting at d_l^1 in directions (u_l^1, d_l^1) and (v_l^1, d_l^1) (the dotted arrows in Fig. 8). They disconnect the grid of lattice points into two 1-connected components. Both components contain vertices of H (e.g. v_l^1 and v_l^2 are in different components). Since H, as a Morpion 5T position, is 1-connected, there must be a vertex of H on at least one of those half-lines. Since d_l^1 does not belong to V_H (as $d_l^1 \in l$ and l is disjoint from V_H as a gap line), at least one of the edges $(u_l^1, d_l^1), (v_l^1, d_l^1)$ belongs to $\mathcal{B}^{\mathrm{int}}(H)$.

Second, we'll show how to pick edges from the edge list $(**_l^i)$ in such a way that the assignment is unique (one-to-one).

Consider edge e_l^1 that crosses a gap line l. There may be only one gap line m such that the edge lists $(**_m^1)$ or $(**_m^2)$ overlap with the edge list $(**_l^1)$. We consider four cases about how edges e_l^2, e_m^1, e_m^2 are placed around e_l^1 (Fig. 9).

Case 1. If both e_m^1 and e_m^2 are vertex disjoint from e_l^1, then we assign to e_l^1 any edge from $(**_l^1)$ that belongs to $\mathcal{B}^{\mathrm{int}}(H)$.

Case 2. Exactly one of e_m^1 and e_m^2 has a common vertex with e_l^1. Without a loss of generality we may assume that e_m^1 has a common vertex $u_l^1 = u_m^1$ with e_l^1. We must be careful to not assign edge (u_l^1, d_l^1) to both edges e_l^1 and e_m^1. We assign (v_l^1, d_l^1) to e_l^1 and (v_m^1, d_m^1) to e_m^1.

Case 3. Both e_m^1 and e_m^2 have a common vertex with e_l^1 but e_l^2 is vertex disjoint from e_m^1 and e_m^2. Assume that v_m^1 and u_m^2 are vertices of e_m^1 and e_m^2 that are disjoint from e_l^i. We assign (v_m^1, d_m^1) to e_m^1, (u_m^2, d_m^2) to e_m^2 and (u_l^1, d_l^1) to e_l^1.

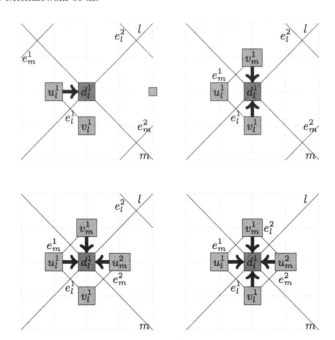

Fig. 9. Two gap lines l and m intersecting at a point d_l^1. Four figures relate to four cases in the proof. The top left is related to Case 1, the top right to Case 2, the bottom left to Case 3 and the bottom right to Case 4.

Case 4. Edges e_l^1, e_l^2, e_m^1, e_m^2 form a small "diamond" (they pairwise intersect) with $d_l^1 = d_l^2 = d_m^1 = d_m^2$ inside. Assuming that the vertices are labeled in such a way that u_k^j are disjoint, we assign (u_k^j, d_k^j) to e_k^j.

This concludes the argument that $2 \operatorname{gap}(H) \leq \operatorname{potential}^{\mathrm{int}}(H)$.

We will now show that the singular moves $m_{k+1}, \ldots, m_n$ add at least $2 \cdot (n - k) - \operatorname{modifier}(\operatorname{hull}(G))$ to the internal potential of the position (i.e. $\operatorname{potential}(G) - \operatorname{potential}(H) \geq 2 \cdot (n - k) - \operatorname{modifier}(\operatorname{hull}(G)))$.

First, observe that there are at most 4 singular moves, that is $n - k \leq 4$. This is because there are two diagonal directions and two sides of a line where the initial cross may be.

Assume that move m_i ($i > k$) places a singular vertex v_i. Let l_i^1, l_i^2 be half-lines starting from v_i in the direction of the gap line created by move m_i (the dashed lines in Fig. 10). There are two possibilities.

Case I. At least one of the half-lines l_i^1, l_i^2 contain a vertex of the position graph. Let u_i denote this vertex. If so, then placing of v_i creates two new edges of internal potential (one starting in v_i in the direction of u_i and another one in u_i in the direction of v_i). The new gap line is compensated.

Case II. Neither of the half-lines l_i^1 and l_i^2 contain a vertex of the position graph. Observe that this is possible only for at most two of the singular moves

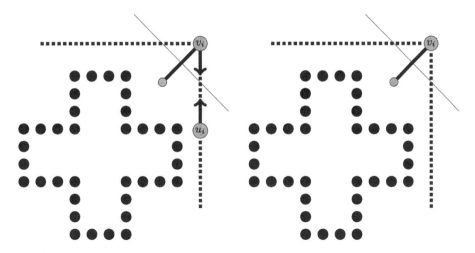

Fig. 10. Two cases appearing in the analysis of the corners. The figure on the left relates to Case I and the figure on the right to Case II.

and each of those moves must create a corner in the hull of G. Moreover, if there are two such moves, the corners are opposite corners of hull(G).

This concludes the proof of the Lemma.

References

1. Boyer, C.: Morpion solitaire (2008). http://www.morpionsolitaire.com
2. Demaine, E.D., Demaine, M.L., Langerman, A., Langerman, S.: Morpion solitaire. Theory Comput. Syst. **39**(3), 439–453 (2006)
3. Gurobi Optimization, I.: Gurobi optimizer reference manual (2015). http://www.gurobi.com
4. Kawamura, A., Okamoto, T., Tatsu, Y., Uno, Y., Yamato, M.: Morpion solitaire 5d: a new upper bound 121 on the maximum score. In: CCCG (2013)
5. Michalewski, H., Nagrko, A., Pawlewicz, J.: Linear programs giving a new upper bound in the Morpion 5T game (2015). https://github.com/anagorko/morpion-lpp
6. Rosin, C.D.: Nested rollout policy adaptation for monte carlo tree search. In: IJCAI, pp. 649–654 (2011)

Multi-agent Retrograde Analysis

Tristan Cazenave[(✉)]

LAMSADE, Université Paris-Dauphine, Paris, France
`cazenave@lamsade.dauphine.fr`

Abstract. We are interested in the optimal solutions to multi-agent planning problems. We use as an example the predator-prey domain which is a classic multi-agent problem. We propose to solve it on small boards using retrograde analysis.

1 Introduction

The predator-prey problem is a classic multi-agent problem. It was introduced in [3]. There are four predators and one prey and the goal of the predators is to capture the prey. In this seminal work the predators can occupy the same location and the prey moves randomly. In a posterior work the agents could not occupy the same location [18]. Richard Korf proposed a simple pursuit strategy using attraction between the predators and the prey and repulsion between predators [10].

The predator-prey problem has been used to test multiple agent based algorithms. For example it has been use to analyze a general model of multi-agent communication with a message board, using a genetic algorithm to evolve multi-agent languages [9]. It has also been used to test genetic algorithms with Lamarckian learning operators in multi-agent environments [7].

Genetic programming has also been used to co-evolve predators and preys populations [8]. In this work the authors acknowledge that the approach fails and claim that a simple prey algorithm is able to evade capture from the predators' algorithms. Another work evolving multi-agent teams for the predator-prey game is presented in [11].

In a system evolving neural networks in separate subpopulations for different agents, it was advocated that the learning is easier than with a single controller and that communication is unnecessary and even detrimental in the predator-prey problem [22].

Another line of work is the development of almost optimal algorithms for the prey [12]. These algorithms were tested on twenty maps of the commercial computer game Baldur's Gate and the best one achieved up to 98 % optimality, while being reasonably fast.

The other topic we address in this paper is retrograde analysis. Retrograde analysis computes the optimal solution to a large number of game states starting from terminal positions and going up towards deeper positions. It was first used in Chess and Checkers and related to dynamic programming [1].

© Springer International Publishing Switzerland 2016
T. Cazenave et al. (Eds.): CGW 2015/GIGA 2015, CCIS 614, pp. 60–70, 2016.
DOI: 10.1007/978-3-319-39402-2_5

Chess endgames have been completely solved up to 6 pieces using retrograde analysis [19,20]. The 6-piece endgame table requires 1.2 TB. Endgame tables have also been instrumental in solving Checkers [17]. Awari was completely solved thanks to retrograde analysis [13] leading to an optimal and instantaneous player. Retrograde analysis has also been applied to many other games such as Nine Men's Morris [6], Go [4], Fanorona [15], Chinese Chess [5] or Chinese Dark Chess [14] among others.

Games can also be solved by search. A standard algorithm for solving games is iterative deepening $\alpha\beta$ with a transposition table [16]. Search was used to solve Checkers [17], small board Go [21] and small board Atari Go [2].

The outline of the paper is to present the predator-prey game in the next section, then to present retrograde analysis and search, followed by experimental results.

2   The Predator-Prey Game

In the predator-prey game we have designed, three predators are trying to capture a prey. In our implementation there are five possible moves for each agent: going up, down, left, right or staying on the same location. Predators cannot occupy the same location and when a prey moves to a predator location it is captured.

A state is terminal either if the prey is on the same location as a predator or if the prey is blocked by the predators and cannot move to an empty location.

A state is legal if no two predators are on the same location.

In previous work, moves by the predators and the prey can be either simultaneous or sequential. We have chosen sequential moves with the prey moving after the predators. When the prey is the second player he can choose the move the most beneficial to him knowing the future locations of the predators. It should be better for the prey but the evaluation must be made only after the prey move so as to simulate simultaneous moves. If the predator moves to the location of the prey, the prey can still escape since it can still move and that the evaluation of a state is only made after the move of the prey.

In our implementation it is possible for the prey to swap locations with a neighbor predator. It could also be possible to forbid such swaps. Enabling swaps as we do should be beneficial to the prey.

Overall when we had to make choices for the design of the game we chose the design the most beneficial to the prey.

3   Retrograde Analysis

In order to store the results of retrograde analysis in a table we have to design a bijection between the states of the problem and the indices in the table. We call the index associated to a state its code. A simple way to compute a code is to number each agent and each cell on the board and to compute the code of a board as:

$$code = \sum\nolimits_{agent} cell(agent) \times MaxCell^{agent}$$

In this formula the agent variable is an integer between 0 and 3 that represents an agent. Agent 0 is the prey and agents 1, 2 and 3 are the predators. The function $cell(agent)$ returns an integer that represents the location of the agent on the board, each cell is associated to an integer between 0 and $MaxCell - 1$.

Using the previous code we consider that each agent is different from the other ones. However we could consider that two predators can exchange their locations and that it is still the same state. In this case the total number of states and the greatest possible code are quite reduced [15], thus reducing the size of the retrograde analysis table.

For this paper, we kept things simple and in our experiments we used the simple code considering each agent different from the other ones.

The overall algorithm that performs retrograde analysis is given in Algorithm 4. It calls two subsequent algorithms. The initialization algorithm that is given in Algorithm 1 and which initializes the table with terminal states, and the step algorithm that is given in Algorithm 2 and which computes the states won by the predators in currentDepth moves by the prey. The step algorithm performs a one ply search in order to discover the states won at currentDepth given the states won in less than currentDepth. This one-ply search algorithm is given in Algorithm 3.

In the algorithms the constant $MaxAgent$ is set to 4 and represents the number of agents including the prey. When the $agent$ variables reaches $MaxAgent$ it means that either all the agents have been placed in the initialization algorithm or that all the predators have moved in the step and the one step lookahead algorithms.

In Algorithm 3 the predators try to minimize the depth to the capture and the prey tries to maximize it. The unknown states are initialized to ∞ in the init algorithm. If the prey can escape to an unknown state then the algorithm returns ∞ and the predators have to keep trying other moves.

The table can be used to decide the predators' moves that win in the smallest number of steps. It can also be used to decide the prey move that will take the most number of steps before capture. In some Chess endgames, even in lost states, computers using an endgame table can lure grandmasters and keep them away from victory as the human players do not always play optimal moves.

4   Search

The worst branching factor for 4 agents and vertical and horizontal moves is $5^4 = 625$. A simple depth 8 problem for size 5×5 can already visit at most $625^8 = 2.33 \times 10^{22}$ leaves. In practice the exact number of leaves should be less but still quite a large number. It would be clearly more than the state space size of a 5×5 problem which is 345,000. The state space complexity of the problem is far lower than its game tree complexity.

Algorithm 1. The initialization algorithm

init (*agent*)
if *agent* = MaxAgents **then**
  if board is legal **then**
    nbStates ← nbStates + 1
    depth [board.code ()] ← ∞
    if board is terminal **then**
      depth [board.code ()] ← 0
      nbStatesDepth [0] ← nbStatesDepth [0] + 1
    end if
  end if
else
  for *cell* in possible locations on board **do**
    board.cell [*agent*] ← *cell*
    init (*agent* + 1)
  end for
end if

Algorithm 2. The step algorithm computing the next depth of retrograde analysis

step (*agent*)
if *agent* = MaxAgents **then**
  if depth [board.code ()] = ∞ **then**
    if board is legal **then**
      if min (1) = currentDepth - 1 **then**
        depth [board.code ()] ← currentDepth
        nbStatesDepth [currentDepth] ← nbStatesDepth [currentDepth] + 1
      end if
    end if
  end if
else
  for *cell* in possible locations on board **do**
    board.cell [*agent*] ← *cell*
    step (*agent* + 1)
  end for
end if

A possible solution to avoid searching again the already visited states is to use iterative deepening search with a transposition table as a search algorithm. It avoids searching again the same state multiple times and it could significantly decrease the search time as there are many transpositions in the predator-prey problem.

We have implemented a perfect transposition table. A perfect transposition table is a table that has exactly one entry per possible state. When a state has been searched the result can be stored in the corresponding entry and it can be reused when reaching the state again. We use the code of the board as the index in the transposition table.

Algorithm 3. The one step lookahead algorithm

---

min (*agent*)
if *agent* = MaxAgents **then**
  return max ()
end if
mini ← min (*agent* + 1)
for *move* in possible moves for *agent* **do**
  make *move* for *agent*
  eval ← min (*agent* + 1)
  undo *move* for *agent*
  if *eval* < *mini* **then**
    mini ← *eval*
  end if
end for
return *mini*

max ()
if board is illegal **then**
  return ∞
end if
maxi ← depth [board.code ()]
for *move* in possible moves for the prey **do**
  make *move* for the prey
  eval ← depth [board.code ()]
  undo *move* for the prey
  if *eval* > *maxi* **then**
    maxi ← *eval*
  end if
end for
return *maxi*

Algorithm 4. The overall algorithm for retrograde analysis

---

nbStates ← 0
nbStatesDepth [0] ← 0
init (0)
currentDepth ← 1
while true **do**
  nbStatesDepth [currentDepth] ← 0
  step (0)
  if nbStatesDepth [currentDepth] = 0 **then**
    break
  end if
  currentDepth ← currentDepth + 1
end while

The search algorithm for the predator-prey game is given in Algorithm 5. It uses a perfect transposition table and two functions. The minTT function tries all the possible combinations of the predators' moves and selects the one leading to the capture of the prey if it exists. If no combination enables the capture in *depth* steps it returns false. The maxTT function tries all possible moves for the prey and selects the one that avoids capture. If all possible moves lead to capture it stores the result in the transposition table and returns true.

The TT table contains the depth of the search that solved the state. It contains ∞ if the state was not solved. If a state has already been solved with a smaller or equal depth, the algorithm returns true. The other table is the depthTT table, it contains the maximum search depth performed for the state. If a state has already been searched with a greater or equal depth, the search is cut as it is not necessary to search it again.

5 Experimental Results

The experiments were run on a 1.9 GHz computer running Linux and the algorithms were written in C++.

The retrograde analysis algorithm was used to compute the depth to mate of every state for various board sizes. The number of states for each depth to mate is given in Table 1 for board sizes ranging from 4×4 to 9×9.

Table 2 gives the total number of states, the maximum code used and the time to perform retrograde analysis for 4×4 to 9×9 boards.

We wrote an algorithm similar to the initialization algorithm in order to verify that all states are won for the predators. It is run after the retrograde analysis is finished and verifies that the depth to mate is finite for every possible state. We have found that it is the case for all the board sizes we solved.

A 7×7 state with maximum depth 12 is the following state:

```
o....xx
......x
.......
.......
.......
.......
.......
```

The iterative deepening search for this state evolves as indicated in Table 3. The times indicated are the cumulative times, the times for all inferior depth. The time to solve the corresponding 9×9 problem with search at depth 16 is 1714.53 s. This is only to solve one problem when retrograde analysis can be computed offline in 1,900 s for size 7×7 and for all problems and results in instantaneous and optimal moves.

In order to illustrate the predators' strategies, we give an example of the solution to the 7×7 problem above of maximum depth 12, with the prey randomly choosing among the moves leading to a maximum depth state:

Algorithm 5. The search algorithm

minTT (*depth, agent*)
if *agent* = MaxAgents then
  return maxTT (*depth* − 1)
end if
if minTT (*depth, agent* + 1) then
  return true
end if
for *move* in possible moves for *agent* do
  make *move* for *agent*
  eval ← minTT (*depth, agent* + 1)
  undo *move* for *agent*
  if *eval* = *true* then
    return true
  end if
end for
return false

maxTT (*depth*)
if board is illegal then
  return false
end if
if prey is blocked then
  return true
end if
if *depth* = 0 then
  return false
end if
if TT [board.code ()] ≤ *depth* then
  return true
end if
if depthTT [board.code ()] ≥ *depth* then
  return false
end if
if depthTT [board.code ()] < *depth* then
  depthTT [board.code ()] = *depth*
end if
if the prey is not on the same location as a predator then
  if not minTT (*depth*, 1) then
    return false
  end if
end if
for *move* in possible moves for the prey do
  make *move* for the prey
  if the prey is not on the same location as a predator then
    if not minTT (*depth*, 1) then
      undo *move* for the prey
      return false
    end if
  end if
  undo *move* for the prey
end for
TT [board.code ()] ← *depth*
return true

Table 1. Number of states for each depth to mate and for different board sizes.

Depth	4×4	5×5	6×6	7×7	8×8	9×9
0	10,440	42,000	129,408	332,856	751,560	1,537,800
1	2,712	4,920	7,464	10,344	13,560	17,112
2	3,960	5,976	7,560	8,616	9,672	10,728
3	10,200	16,056	20,808	24,792	26,760	28,728
4	19,584	42,840	48,960	57,888	64,752	68,352
5	6,864	79,752	119,040	128,616	138,912	150,912
6	0	83,928	240,864	273,600	283,416	298,080
7	0	68,760	367,896	531,000	584,280	580,128
8	0	768	387,816	888,696	1,122,336	1,131,336
9	0	0	211,848	1,218,600	1,927,536	2,067,480
10	0	0	576	1,170,576	3,021,264	3,533,112
11	0	0	0	755,424	3,478,080	5,575,200
12	0	0	0	15,648	3,005,280	7,666,608
13	0	0	0	0	1,551,816	8,301,696
14	0	0	0	0	19,752	6,625,848
15	0	0	0	0	0	3,816,432
16	0	0	0	0	0	55,968
17	0	0	0	0	0	0

```
O....XX    O....X.    O....X.    ....X..    ...X...    ..X....
.....X     ....XX     ...XX.     O..XX..    ..XX...    .XX....
.......    .......    .......    .......    O......    .......
.......    .......    .......    .......    .......    O......
.......    .......    .......    .......    .......    .......
.......    .......    .......    .......    .......    .......
.......    .......    .......    .......    .......    .......
  12         11         10         9          8          7


.X.....    .......    .......    .......    .......    .......
.X.....    .X.....    .......    .......    .......    .......
..X....    .X.....    .X.....    .......    .......    .......
.O.....    .OX....    .XX....    .X.....    X......    X......
.......    .......    .O.....    OXX....    XX.....    .......
.......    .......    .......    .......    O......    XX.....
.......    .......    .......    .......    .......    O......
   6          5          4          3          2          1
```

Table 2. Number of states, maximum code and time to solve with retrograde analysis in seconds for different sizes

Size	Number of states	Maximum code	Time to solve
4×4	$53,760$	$65,536$	3.82
5×5	$345,000$	$390,625$	56.16
6×6	$1,542,240$	$1,679,616$	386.75
7×7	$5,416,656$	$5,764,801$	$1,900.91$
8×8	$15,998,976$	$16,777,216$	$8,618.76$
9×9	$41,465,520$	$43,046,721$	$27,002.54$

Table 3. Times for searching a 7×7 depth 12 state with iterative deepening and a perfect transposition table.

Depth	Time
1	0.006094
2	0.006181
3	0.007001
4	0.013914
5	0.046882
6	0.162278
7	0.492648
8	1.453308
9	4.410059
10	12.724911
11	91.924769
12	164.675883

6   Conclusion

Retrograde analysis of the predator-prey problem is tractable in time and memory until 9×9 boards. It results in instantaneous decisions and optimal multi-agent strategies.

A result from this research is that the predator-prey game is always lost for the prey even when there are only 3 predators, when the prey knows the moves of the predators before moving and when the prey is allowed to swap locations with a neighbor predator. The maximum number of moves by the prey before capture is 14 for size 8×8 and 16 for size 9×9.

Another result is that iterative deepening search with a perfect transposition table is slow even for small board sizes. It cannot compete with retrograde analysis with respect to solving time.

In future work we will explore the use of abstraction so as to solve boards of large sizes, learning of agents strategies, and compression of tables. Another line of research is to solve a continuous version of the game.

There are multiple possibilities for learning using endgame tables. For example, learning an evaluation function for a depth one search, learning the move to make for an agent or learning a move ordering heuristic with an evaluation of states or with an evaluation of moves.

Another line of research is to analyze endgames of multi-agent games with a much larger state space.

References

1. Bellman, R.: On the application of dynamic programing to the determination of optimal play in chess and checkers. Proc. Nat. Acad. Sci. U.S.A. **53**(2), 244 (1965)
2. Boissac, F., Cazenave, T.: De nouvelles heuristiques de recherche appliquées à la résolution d'Atarigo. In: Intelligence Artificielle et Jeux, pp. 127–141. Hermes Science (2006)
3. Brenda, M., Jagannathan, V., Dodhiawala, R.: On optimal cooperation of knowledge sources-an empirical investigation. Boeing Adv. Technol. Center, Boeing Comput. Services, Seattle, WA, Technical report, BCSG2010-28 (1986)
4. Cazenave, T.: Generation of patterns with external conditions for the game of Go. Adv. Comput. Games **9**, 275–293 (2001)
5. Fang, H., Hsu, T., Hsu, S.-C.: Construction of Chinese Chess endgame databases by retrograde analysis. In: Marsland, T., Frank, I. (eds.) CG 2001. LNCS, vol. 2063, pp. 96–114. Springer, Heidelberg (2002)
6. Gasser, R.: Solving Nine Men's Morris. Comput. Intell. **12**(1), 24–41 (1996)
7. Grefenstette, J.J.: Lamarckian learning in multi-agent environments. Technical report, DTIC Document (1995)
8. Haynes, T., Sen, S.: Evolving behavioral strategies in predators and prey. In: Weiss, G., Sen, S. (eds.) IJCAI-WS 1995. LNCS, vol. 1042, pp. 113–126. Springer, Heidelberg (1996)
9. Jim, K.C., Giles, C.L.: Talking helps: evolving communicating agents for the predator-prey pursuit problem. Artif. Life **6**(3), 237–254 (2000)
10. Korf, R.E.: A simple solution to pursuit games. In: Proceedings of the 11th International Workshop on Distributed Artificial Intelligence, pp. 183–194 (1992)
11. Luke, S., Spector, L.: Evolving teamwork and coordination with genetic programming. In: Proceedings of the 1st Annual Conference on Genetic Programming, pp. 150–156. MIT Press (1996)
12. Moldenhauer, C., Sturtevant, N.R.: Evaluating strategies for running from the cops. In: Proceedings of the 21st International Joint Conference on Artificial Intelligence, IJCAI 2009, Pasadena, California, USA, 11–17 July 2009, pp. 584–589 (2009)
13. Romein, J.W., Bal, H.E.: Solving Awari with parallel retrograde analysis. IEEE Comput. **36**(10), 26–33 (2003)
14. Saffidine, A., Jouandeau, N., Buron, C., Cazenave, T.: Material symmetry to partition endgame tables. In: Herik, H.J., Iida, H., Plaat, A. (eds.) CG 2013. LNCS, vol. 8427, pp. 187–198. Springer, Heidelberg (2014)
15. Schadd, M.P.D., Winands, M.H.M., Uiterwijk, J.W.H.M., van den Herik, H.J., Bergsma, M.H.J.: Best play in Fanorona leads to draw. New Math. Nat. Comput. **4**(3), 369–387 (2008)

16. Schaeffer, J.: The history heuristic and alpha-beta search enhancements in practice. IEEE Trans. Pattern Anal. Mach. Intell. **11**(11), 1203–1212 (1989)

17. Schaeffer, J., Burch, N., Björnsson, Y., Kishimoto, A., Müller, M., Lake, R., Lu, P., Sutphen, S.: Checkers is solved. Science **317**(5844), 1518–1522 (2007)

18. Stephens, L.M., Merx, M.B.: Agent organization as an effector of DAI system performance. In: Ninth Workshop on Distributed Artificial Intelligence, Rosario Resort, Eastsound, Washington, pp. 263–292 (1989)

19. Thompson, K.: Retrograde analysis of certain endgames. ICCA J. **9**(3), 131–139 (1986)

20. Thompson, K.: 6-piece endgames. ICCA J. **19**(4), 215–226 (1996)

21. van der Werf, E.C.D., Winands, M.H.M.: Solving Go for rectangular boards. ICGA J. **32**(2), 77–88 (2009)

22. Yong, C.H., Miikkulainen, R.: Cooperative coevolution of multi-agent systems. Technical report, AI07-338, University of Texas at Austin, Austin, TX (2001)

The Surakarta Bot Revealed

Mark H.M. Winands[(✉)]

Games and AI Group, Department of Data Science and Knowledge Engineering,
Maastricht University, Maastricht, The Netherlands
`m.winands@maastrichtuniversity.nl`

Abstract. The board game Surakarta has been played at the ICGA
Computer Olympiad since 2007. In this paper the ideas behind the
agent SIA, which won the competition five times, are revealed. The
paper describes its $\alpha\beta$-based variable-depth search mechanism. Search
enhancements such as multi-cut forward pruning and Realization Probability Search are shown to improve the agent considerably. Additionally,
features of the static evaluation function are presented. Experimental
results indicate that features, which reward distribution of the pieces
and penalize pieces that clutter together, give a genuine improvement in
the playing strength.

1 Introduction

Since 2007 the board game Surakarta has been played six times at the ICGA
Computer Olympiad, a multi-games event in which all of the participants are
computer programs. The Surakarta agent SIA won the gold medal at the 12th,
13th, 15th, 17th, and 18th ICGA Computer Olympiad. It did not lose a single
game in each tournament it participated.

In this paper the $\alpha\beta$-search based agent SIA is discussed in detail. It presents
SIA's variable-depth search mechanism [9] that contains quiescence search [12],
multi-cut forward pruning [2] and Realization Probability Search [13]. Also, the
features of the static evaluation function are described and assessed.

The article is organized as follows. First, in Sect. 2 the game of Surakarta is
briefly discussed. Next, SIA's $\alpha\beta$-search engine is introduced in Sect. 3. In Sect. 4
its variable-depth search mechanism is described. Subsequently, the evaluation
function is proposed in Sect. 5. The experimental results are presented in Sect. 6.
Finally, Sect. 7 gives conclusions and an outlook on future research.

2 Surakarta

Surakarta is a board game for two players (i.e., Black and White). It is played on a
6×6 board where eight loops extend out from it (see Fig. 1). The four small loops
form together the inner circuit, whereas the four large loops form the outer circuit.

Players take turns moving one of their own pieces. In non-capturing moves, a
piece travels – either orthogonally or diagonally – to a neighboring intersection.

© Springer International Publishing Switzerland 2016
T. Cazenave et al. (Eds.): CGW 2015/GIGA 2015, CCIS 614, pp. 71–82, 2016.
DOI: 10.1007/978-3-319-39402-2_6

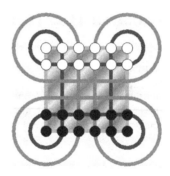

Fig. 1. Initial Surakarta position.

In a capturing move, a piece travels along a line, *traveling over at least one loop*, until it meets one of the opponent pieces. The captured piece is removed, and the capturing piece takes its place. The first player to capture all opponent's pieces wins. Draws can occur by repetition of moves or stalemate (cf. [6]). In this article, if a position with the same player to move occurs for the third time, the game is drawn. Additionally, if in the last fifty moves no capture was made, the game is scored as a draw as well.

Self-play experiments by SIA revealed that the game has an average branching factor of approximately 22 and an average game length of around 54 ply. The game-tree complexity is estimated to be about 10^{72}. Taking symmetry into account, its state-space complexity is 10^{15}.

3   SIA

SIA performs an $\alpha\beta$ depth-first iterative-deepening search in the PVS framework [10]. A *two-deep* transposition table [3] is applied to prune a subtree or to narrow the $\alpha\beta$ window. At all interior nodes that are more than 2 ply away from the leaves, it generates all moves to perform Enhanced Transposition Cutoffs (ETC) [11]. For move ordering, the move stored in the transposition table (if applicable) is always tried first, followed by two killer moves [1]. These are the last two moves that were best, or at least caused a cutoff, at the given depth. Thereafter follow the capture moves. All the remaining moves are ordered decreasingly according to the relative history heuristic [16].

4   Variable-Depth Search

The $\alpha\beta$ algorithm [8] is still the standard search procedure for playing material-based board games such as chess and checkers. The playing strength of programs employing $\alpha\beta$ search depends greatly on how deep they search critical lines of play. Therefore, over the years, many techniques for augmenting $\alpha\beta$ search with a more selective tree-expansion mechanism have been developed, so called *variable-depth search* techniques [9]. Promising lines of play are explored more deeply (search extensions), at the cost of other less interesting ones that are cut off prematurely (search reductions or forward pruning).

In the Surakarta engine SIA the following techniques are employed: *quiescence search* [7,12], *multi-cut* [2], and *Realization Probability Search (RPS)* [13]. They are described in Subsects. 4.1, 4.2, and 4.3, respectively.

4.1   Quiescence Search

When the $\alpha\beta$ search reaches the depth limit, a static evaluation function should be applied in the leaf node reached. This approach can have disastrous consequences because of the approximate nature of the evaluation function. Therefore a more sophisticated cut-off may be required. The evaluation function should only be applied to positions that are *quiescent*.

At the leaf nodes of the regular search, a quiescence search is performed to get more accurate evaluations. In SIA an extended version of quiescence search is implemented [12]. This type of a quiescence search limits the set of moves to be considered and uses the evaluations of interior nodes as lower/upper bounds of the resulting search value. As capture moves are responsible for swings in the evaluation function in Surakarta, only captures are considered for this part of the search.

4.2   Multi-cut

Multi-cut pruning is a forward-pruning technique [2], which has been applied in chess and Lines of Action [15]. Before examining a node to full depth, the first M child nodes are searched to a depth reduced with a factor R. If at least C child nodes return a value larger than or equal to β, a cutoff occurs. However, if the pruning condition is not satisfied, the search continues as usual, re-exploring the node under consideration to a full depth d. In general the behavior of multi-cut is as follows. The higher M and R are and the lower C is, the higher the number of prunings is.

An enhanced version of multi-cut [15] is used in SIA. First, when at a reduced depth a winning value is found, the search is stopped and the winning value is returned. Second, if the multi-cut does not succeed in causing a cutoff, the moves causing a β-cutoff at the reduced depth are tried first in the normal search. Third, multi-cut is used in all nodes, except in the expected principal variation (so-called PV nodes). The idea is that it is too risky to prune forward there, because a possible mistake causes an immediate change of the principal variation. For all other nodes (so-called CUT and ALL nodes [9]), multi-cut is performed with the following parameter settings: $C = 3$ for a CUT node, $C = 2$ for an ALL node, $M = 10$ and $R = 2$ for both node types. The pseudo code in the PVS framework is given in Fig. 2.

```
//Forward-pruning code
 if(node.node_type != PV_NODE && depth > 2){
   next = firstSuccessor(node);
   c = 0, m = 0;
   while(next != null && m < M){
     value = -PVS(next, -beta, -alpha, depth-1-R);
     if(value >= beta){
       c++;
       //Keep track of the moves causing a cut-off at d-R
       storeCutOffNode(next);
       if(value >= WIN_SCORE)
          return value;
       else if(c >= C)
          return beta;
     }
     m++;
     next = nextSibling(next);
   }
   //Re-order moves
   putCutOffNodesInFront();
 }
```

Fig. 2. Pseudo code for multi-cut

4.3 Realization Probability Search

One successful member of the family of variable-depth search techniques is *Realization Probability Search (RPS)*, introduced by Tsuruoka *et al.* [13] in 2002. Using this technique his program, GEKISASHI, won the 2002 World Computer Shogi Championship, resulting in the algorithm gaining a wide acceptance in computer Shogi. It has been successfully applied in the Lines-of-Action engine MIA as well [14].

The RPS algorithm is an approach of using fractional-ply extensions. The algorithm uses a probability-based approach to assign fractional-ply weights to move categories, and then uses re-searches to verify selected search results.

First, for each move category one must determine the probability that a move belonging to that category will be played. This probability is called the *transition probability*. This statistic is obtained from game records of matches played by expert players. The transition probability for a move category c is calculated as follows:

$$P_c \leftarrow \frac{n_{played(c)}}{n_{available(c)}} \tag{1}$$

where $n_{played(c)}$ is the number of game positions in which a move belonging to category c was played, and $n_{available(c)}$ is the number of positions in which moves belonging to category c were available.

Originally, the *realization probability* of a node represented the probability that the moves leading to the node will be actually played. By definition, the realization probability of the root node is 1. The transition probabilities of moves were then used to compute the realization probability of a node in a recursive manner (by multiplying together the transition probabilities on the path leading to the node). If the realization probability would become smaller than a predefined threshold, the node would become a leaf. Since a probable move has a large transition probability while an improbable has a small probability, the search proceeds deeper along probable move sequences than improbable ones.

Instead of using the transition probabilities directly, they can be transformed into fractional plies [13]. The fractional ply FP of a move category is calculated by taking the logarithm of the transition probability in the following way:

$$FP \leftarrow log_K(P_c) \tag{2}$$

where K is a constant between 0 and 1. A value of 0.25 is a good setting for K in Surakarta. Note that this setting is probably domain dependent, and a different value could be more appropriate in a different game or even game engine.

The fractional-ply values are calculated off-line for all the different move categories, and used on-line by the search (as shown in Fig. 3 [14]). In the case where FP is larger than 1 it means the search is reduced while in the case FP is smaller than 1 the search is extended. By converting the transition probabilities to fractional plies, move weights now get added together instead of being multiplied. This has the advantage that RPS is used alongside multi-cut, which measures depth similarly.

However, setting the depth of the move based on its FP values runs into difficulties because of the horizon effect. Move sequences with high FP values (i.e., low transition probability) get terminated quickly. Thus, if a player experiences a significant drop in its positional score as returned by the search, it is eager to play a possibly inferior move with a higher FP value, simply to push the inevitable score drop beyond its search horizon.

To avoid this problem, RPS is instructed to perform a *deeper* re-search for a move whose value is larger than the current best value (i.e., the α value). Instead of reducing the depth of the re-search by the fractional-ply value of the move (as is generally done), the search depth is decreased only by a small predefined FP value, called *minFP*. It is set equal to the lowest move category value.

Apart from how the ply depth is determined, and the re-search, the algorithm is otherwise almost identical to PVS [10]. Figure 4 shows a C-like pseudo-code. Because the purpose of the preliminary search is only to check whether a move will improve upon the current best value, a null-window may be used.

RPS is applied in SIA in the following way. First, moves are classified as captures or non-captures. Next, moves are further subclassified based on the origin and destination of the move's from and to squares. The board is divided into four different regions: the corners, the 6×6 outer rim (except corners), the 4×4 inner rim, and the central 2×2 board. In total 20 move categories can occur in the game according to this classification. The transition probabilities have been collected by letting SIA play 1000 games against itself. The final FP

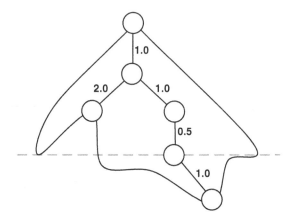

Fig. 3. Fractional-ply example for a nominal search depth of 3 [14].

values of the move categories are capped between 0.5 and 4.0 (inclusive). They are shown in Table 1.

When looking at the transition probabilities, capture moves are in general preferred above non-capture moves. Although moving away from a corner is also strongly encouraged. Interestingly, when a move is a non-capture it is better to move towards the center. In case of a capture move, the opposite is true.

5   Evaluation Function

In this section the relevant features of the static evaluation function are enumerated and explained. The evaluator consists of the following five features: *material, mobility, player to move, quads,* and *distribution*. The choice of features that fully cover the description of a position is most relevant. It is better to have all features correct and all the initial weights wrong than to have the initial weights correct and miss one of the (important) features. The description of the features follows below; relevant examples and clarifications are given, adequate references to further details are supplied. It is followed by some information about the use of caching.

Material. Analogous to piece-square tables in chess, each piece obtains a value dependent on its board square in SIA. Especially, pieces at the corner are evaluated less. The relative values are given in the following matrix:

$$
\begin{bmatrix}
3 & 10 & 10 & 10 & 10 & 3 \\
10 & 11 & 10 & 10 & 11 & 10 \\
10 & 10 & 10 & 10 & 10 & 10 \\
10 & 10 & 10 & 10 & 10 & 10 \\
10 & 11 & 10 & 10 & 11 & 10 \\
3 & 10 & 10 & 10 & 10 & 3
\end{bmatrix}
$$

```
RPS(node, alpha, beta, depth){
    //Transposition table lookup, omitted
    ......................................
    if(depth <= 0)
      return quiescenceSearch(node, alpha, beta);
    //Do not perform forward pruning in a potential principal variation
    if(node.node_type != PV_NODE){
      //Multi-cut code, omitted
      ......................................
      if(forward_pruning condition holds) return beta;
    }
    next = firstSuccessor(node);
    while(next != null){
      alpha = max(alpha, best);
      decDepth = FP(next);
      //Preliminary Search Null-Window Search Part
      value = -RPS(next, -alpha-1, -alpha, depth-decDepth);
      //Re-search
      if(value > alpha)
        value = -RPS(next, -beta, -alpha, depth-minFP);

      if(value > best){
        best = value;
        if(best >= beta) goto Done;
      }
      next = nextSibling(next);
    }

    Done: //Store in Transposition table, omitted
    ......................................
}
```

Fig. 4. Pseudo code for Realization Probability Search.

Mobility. Having more moves than the opponent may imply that you have more "freedom" that can be correlated with success. The computational requirements of the mobility feature are not high if only non-capture moves are considered. For each line configuration (represented as a bit vector) the mobility can be precomputed and stored in a table. During the search, the index scheme can be updated incrementally and in the evaluation function only a few table lookups have to be done.

An advantage of this feature that it is fast to evaluate. A disadvantage of this implementation is that capture moves are not taken into account. This is *partially* mitigated by the quiescence search as only leaf nodes are evaluated that cannot start a capture sequence anymore. Still, it could be that the *non-moving* player has several possibilities to capture. Quiescence search is therefore not able to completely assess the capturing potential of one of the players.

Table 1. Move categories together with their transition probabilities and *FP* values.

Capture	Destination	Target	Transition Probability	FP value
No	Corner	Outer Rim	30.4 %	0.85
No	Corner	Inner Rim	48.4 %	0.52
No	Outer Rim	Corner	1.6 %	2.97
No	Outer Rim	Outer Rim	12.9 %	1.47
No	Outer Rim	Inner Rim	17.0 %	1.27
No	Inner Rim	Corner	0.8 %	3.45
No	Inner Rim	Outer Rim	6.7 %	1.94
No	Inner Rim	Inner Rim	6.7 %	1.95
No	Inner Rim	Center	11.5 %	1.55
No	Center	Inner Rim	2.7 %	2.60
No	Center	Center	7.4 %	1.88
Yes	Outer Rim	Outer Rim	64.3 %	0.50
Yes	Outer Rim	Inner Rim	59.0 %	0.50
Yes	Outer Rim	Center	51.9 %	0.50
Yes	Inner Rim	Outer Rim	63.4 %	0.50
Yes	Inner Rim	Inner Rim	58.6 %	0.50
Yes	Inner Rim	Center	49.4 %	0.50
Yes	Center	Outer Rim	50.9 %	0.50
Yes	Center	Inner Rim	47.2 %	0.54
Yes	Center	Center	42.7 %	0.61

Player to Move. The player-to-move feature is based on the basic principle of the initiative. It rewards the moving side. Having the initiative is mostly an advantage in Surakarta like in many other games.

Since SIA is using variable-depth search (because of quiescence search, the multi-cut, and RPS) not all leaf nodes are evaluated at the same depth. Therefore, leaf nodes in the search tree may have a different player to move, which is compensated in the evaluation function. This is done by giving a small bonus to the side to move.

Distribution. The distribution feature is based on the principle of spreading the pieces over the board to increase the potential to attack pieces of the opponent. In SIA this is done in a way which is primitive but effective. First the maximum number m of pieces of a player in a row or column is determined. The distribution is calculated as follows:

$$distribution = \frac{25 \times n}{max(2, m)} \qquad (3)$$

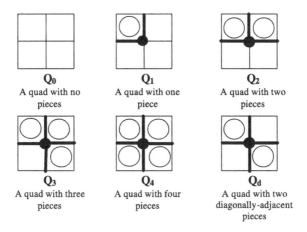

Fig. 5. Six different quad types.

where n is the number of pieces of a player. In such a way this feature prevents that there are too many pieces on one line. It is connected to the following feature, *quads*, that penalizes solid formations.

Quads. The quads feature prevents that pieces are cluttered together. The heuristic is based on the use of quads, an Optical Character Recognition method. A quad is defined as a 2×2 array of squares [5]. Taking into account rotational equivalence, there are six different quad types, depicted in Fig. 5. The values of each quad type is given in Table 2. Quads with 1 or 2 pieces receive a bonus, whereas quads with 4 pieces get a penalty.

Table 2. Quad values.

Quad types	Q_1	Q_2	Q_3	Q_4	Q_d
Values	5	5	0	−5	10

Caching Features. It is possible in SIA's evaluation function to cache computations of certain features, which can be used in other positions. The material, quads, and distribution features are independent of the position of the other side. They are stored in an evaluation cache table. In the current evaluation function this gives a speed-up of at least 30 % in the number of nodes investigated per second.

6 Experiments

In this section the main components of SIA are tested. Different versions of SIA played at least 1000 games against each other, playing both colors equally.

To prevent that games were repeated, a random factor was included in the evaluation function. Draws were considered half wins to each player to ensure the winning percentages sum to 100 %. All experiments were performed on an Intel Xeon 5355 2.66 GHz computer. The engine has been implemented in Java. The remainder of this section is organized as follows. First, the variable-depth search techniques are tested in Subsect. 6.1. Next, the features of the evaluation function are assessed in Subsect. 6.2. Finally, SIA's performance on the ICGA Computer Olympiads is briefly discussed in Subsect. 6.3.

6.1 Variable-Depth Search Experiments

In the first series of experiments SIA is instantiated using the various combinations of variable-depth search introduced in Sect. 4. A three-tuple $(RPS, Multi\text{-}Cut, QuiescenceSearch)$ is to represent the parameter setting used in each particular player instance. E.g., for the instantiation $\text{SIA}_{(off, multi, quiescence)}$, RPS is disabled, multi-cut and quiescence search are enabled.

For these experiments, the thinking time was limited to 5 s per move. The variable-depth search techniques were initially tested in an incremental way starting first with quiescence search, adding next multi-cut, and finally incorporating RPS. The first three rows of Table 3 show the results for them. It reveals that every search enhancement makes more or less the same contribution by increasing the winning percentage to approximately 70 % for each addition. In the fourth row it was validated whether multi-cut does give an additional benefit to the RPS framework. By winning 63.5 % of the games multi-cut is a genuine improvement. In the last row the results are given when SIA with all the enhancements played against the default fixed-depth version. All techniques combined lead to a 95 % winning percentage. In the next experiment this combination is used.

6.2 Evaluation Function Results

In the last series of experiments four different evaluation functions competed with each other in a round-robin tournament. They are called MATERIAL, MOBILITY, DISTRIBUTION, and SIA. The MATERIAL evaluator consists out of the piece-square table and a small random factor. The MOBILITY evaluator includes the former and incorporates the mobility and the player-to-move feature. Next, DISTRIBUTION includes the distribution feature. Last, SIA adds the quads feature and represents the evaluation function discussed in Sect. 5. The weights of the features were partially tuned by TD-learning, partially manually. In these experiments, the thinking time was limited to 1 s per move.

The results of the round-robin tournament are given in Table 4. Each match data point represents the result of 1,000 games, with both colors played equally. The table shows that every added feature is a genuine improvement. Spreading the pieces over the board improves the performance of the play as the results of the DISTRIBUTION and SIA evaluators indicate.

Table 3. Winning percentage of testing various combinations of variable-depth search techniques. 95 % confidence intervals are given.

		win %
$\text{SIA}_{(off,off,quiescence)}$	$\text{SIA}_{(off,off,off)}$	73.9 ± 1.5
$\text{SIA}_{(off,multi,quiescence)}$	$\text{SIA}_{(off,off,quiescence)}$	70.2 ± 1.4
$\text{SIA}_{(RPS,multi,quiescence)}$	$\text{SIA}_{(off,multi,quiescence)}$	75.3 ± 2.3
$\text{SIA}_{(RPS,multi,quiescence)}$	$\text{SIA}_{(RPS,off,quiescence)}$	63.5 ± 1.0
$\text{SIA}_{(RPS,multi,quiescence)}$	$\text{SIA}_{(off,off,off)}$	95.2 ± 0.8

Table 4. Winning percentage of testing different evaluation functions. 95 % confidence intervals are given. Each data point is based on a 1000-game match.

	Material	Mobility	Distribution	Sia
Material	-	42.9 ± 3.1	38.2 ± 3.0	32.2 ± 2.9
Mobility	57.1 ± 3.1	-	40.6 ± 3.0	35.8 ± 3.0
Distribution	61.8 ± 3.0	59.4 ± 3.0	-	46.7 ± 3.1
Sia	67.8 ± 2.9	64.2 ± 3.0	53.3 ± 3.1	-

6.3 Computer Olympiad Results

Since 2007 SIA has participated in the Surakarta tournaments at the 12th, 13th, 15th, 17th, and 18th ICGA Computer Olympiad. In the competition each agent receives 30 min of thinking time for the whole game, playing an equal number of games for each color. In these five tournaments SIA played a grand total of 32 games against 7 different opponents, winning all of them. This achievement is a validation of the approach to Surakarta proposed in this paper.

7 Conclusion and Future Research

This paper discussed the main components of the Surakarta agent SIA. Results showed that its variable-depth search mechanism improved the search considerably. Besides the classic quiescence search, multi-cut forward pruning and Realization Probability Search gave a boost in the game playing performance. Next, the evaluation function was described. Beside standard features such as material and mobility, features that helped to spread the pieces over the board gave a genuine increase in performance.

For future research adding a feature to determine who controls a circuit would lead potentially to an increase in playing performance. Next, endgame databases could help to improve the strength of the agent and ultimately help to solve the game. So far all endgame databases up to 8 pieces have been generated. Self-play results reveal that it takes on average 40 ply to reach them, which is too deep for a single search. If a 10-piece database or 12-piece database would be generated, it would take 34 or 30 ply, respectively. Larger databases would need several Terabytes of hard drive. An alternative is to use smaller databases and distribute the search over several cores as in done in Job-Level $\alpha\beta$ search [4].

Acknowledgments. Special thanks go to the anonymous referees whose comments helped to improve this paper.

References

1. Akl, S.G., Newborn, M.M.: The principal continuation and the killer heuristic. In: 1977 ACM Annual Conference Proceedings, pp. 466–473. ACM, New York (1977)
2. Björnsson, Y., Marsland, T.A.: Multi-cut alpha-beta pruning in game-tree search. Theoret. Comput. Sci. **252**(1–2), 177–196 (2001)
3. Breuker, D.M., Uiterwijk, J.W.H.M., van den Herik, H.J.: Replacement schemes and two-level tables. ICCA J. **19**(3), 175–180 (1996)
4. Chen, J.C., Wu, I.C., Tseng, W.J., Lin, B.H., Chang, C.H.: Job-Level Alpha-Beta Search. IEEE Trans. Comput. Intell. AI Games **7**(1), 28–38 (2015)
5. Gray, S.B.: Local properties of binary images in two dimensions. IEEE Trans. Comput. **20**(5), 551–561 (1971)
6. Handscomb, K.: Surakarta. Abstr. Games **4**(1), 8 (2003)
7. Kaindl, H., Horacek, H., Wagner, M.: Selective search versus brute force. ICCA J. **9**(3), 140–145 (1986)
8. Knuth, D.E., Moore, R.W.: An analysis of alpha-beta pruning. Artif. Intell. **6**(4), 293–326 (1975)
9. Marsland, T.A., Björnsson, Y.: Variable-depth search. In: van den Herik, H.J., Monien, B. (eds.) Advances in Computer Games 9, pp. 9–24. Universiteit Maastricht, Maastricht (2001)
10. Marsland, T.: A review of game-tree pruning. ICCA J. **9**(1), 3–19 (1986)
11. Schaeffer, J., Plaat, A.: New advances in alpha-beta searching. In: Proceedings of the 1996 ACM 24th Annual Conference on Computer Science, pp. 124–130. ACM, New York (1996)
12. Schrüfer, G.: A strategic quiescence search. ICCA J. **12**(1), 3–9 (1989)
13. Tsuruoka, Y., Yokoyama, D., Chikayama, T.: Game-tree search algorithm based on realization probability. ICGA J. **25**(3), 132–144 (2002)
14. Winands, M.H.M., Björnsson, Y.: Enhanced realization probability search. New Math. Nat. Comput. **4**(3), 329–342 (2008)
15. Winands, M.H.M., van den Herik, H.J., Uiterwijk, J.W.H.M., van der Werf, E.C.D.: Enhanced forward pruning. Inf. Sci. **175**(4), 315–329 (2005)
16. Winands, M.H.M., van der Werf, E.C.D., van den Herik, H.J., Uiterwijk, J.W.H.M.: The relative history heuristic. In: van den Herik, H.J., Björnsson, Y., Netanyahu, N.S. (eds.) CG 2004. LNCS, vol. 3846, pp. 262–272. Springer, Heidelberg (2006)

Learning to Trade in Strategic Board Games

Heriberto Cuayáhuitl(✉), Simon Keizer, and Oliver Lemon

School of Mathematical and Computer Sciences,
Heriot-Watt University, Edinburgh, UK
hc213@hw.ac.uk

Abstract. Automated agents in multiplayer board games often need to trade resources with their opponents—and trading strategically can lead to higher winning rates. While rule-based agents can be used for such a purpose, here we opt for a data-driven approach based on examples from human players for automatic trading in the game "Settlers of Catan". Our experiments are based on data collected from human players trading in text-based Natural Language. We compare the performance of Bayesian Networks, Conditional Random Fields, and Random Forests in the task of ranking trading offers, and evaluate them both in an offline setting and online while playing the game against a rule-based baseline. Experimental results show that agents trained from data from average human players can outperform rule-based trading behavior, and that the Random Forest model achieves the best results.

1 Introduction

Board games with trading strategies aim not only at entertaining people, but also at training them with trading skills. Popular board games of this kind include Airlines Europe, Crude, Last Will, Settlers of Catan, and Power Grid, among others [22]. While these games can be played between humans, they can also be played between computers and humans. The trading behaviors of computer games are usually based on heuristics or optimization methods. The former include carefully tuned rules, and the latter include methods such as Monte-Carlo Tree Search [27] and Reinforcement Learning [23]. However, their application is not trivial due to the complexity of the games, i.e. due to their large state-action spaces. On the one hand, unique situations in the game can be described by a number of variables (e.g. resources available) so that enumerating them would result in very large state spaces. On the other hand, the action space can also be large due to the wide range of unique negotiations (e.g. givable and receivable resources). While one can aim for optimizing the whole game via compression of the search space, one can also aim for a specialized solution. The latter is the focus of this paper by focusing on learning to trade only, rather than learning to play the whole game. In addition, while previous work has focused on optimizing negotiation strategies [23,27], our proposed approach focuses on learning human-like trading.

The rest of the paper describes our proposed approach based on statistical inference for ranking trading negotiations, i.e. the exchange of some resource(s)

© Springer International Publishing Switzerland 2016
T. Cazenave et al. (Eds.): CGW 2015/GIGA 2015, CCIS 614, pp. 83–95, 2016.
DOI: 10.1007/978-3-319-39402-2_7

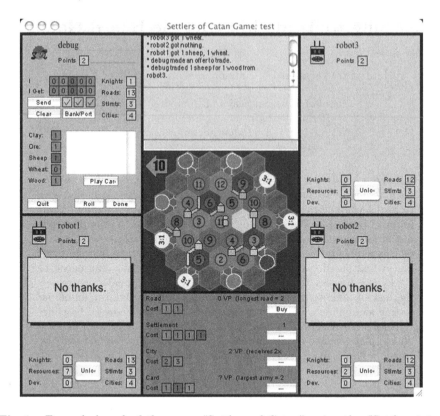

Fig. 1. Example board of the game "Settlers of Catan" using the JSettlers interface [29].

for some other(s). To do that, we use Bayesian networks, Conditional Random Fields, and Random Forests—all trained on data from human examples in the game of Settlers of Catan. We compare our proposed agents against a carefully tuned rule-based agent as a baseline, and show that our learning agents can achieve competitive performance in terms of winning rates.

2   The Game: Settlers of Catan

The game of Settlers of Catan is a multiplayer board game, where players take the role of settlers on the fictitious island of Catan—see Fig. 1. Between two and four players attempt to settle on the island by building settlements and cities connected by roads. To build, players need the following resources: clay, ore, sheep, wheat and wood. Each player gets points for example by building a settlement (1 point) or a city (2 points). A game consists of a sequence of turns, and each game turn starts with the roll of a die that can make the players obtain or lose resources (depending on the number rolled and resources on the board). The player in turn can trade resources with the bank or other players, and can make use of available resources to build roads, settlements or cities. This game is highly strategic because players often face decisions about what resources to request and what resources to give away, which are influenced by what they need to build, for example: a road requires 1 clay and 1 wood. A player can extend build-ups on locations connected to existing pieces, i.e. road, settlement or city, and all settlements and cities must be separated by at least 2 roads. The first player to win 10 victory points wins and all others lose—see the following link for the full set of rules: http://www.catan.com/service/game-rules.

3   The Data and Task

We used a set of 32 logged games from 56 different players, annotated as described in [1]. They correspond to 2512 trading negotiation events (also referred to as 'training instances') denoted as $D = \{(\mathbf{x}_1, y_1), .., (\mathbf{x}_N, y_N)\}$, where $\mathbf{x}_i$ are vectors of features and y_i are class labels (i.e. givable resources). An example trading negotiation in the game of Settlers of Catan in natural language is *"I'll give anyone sheep for clay"*, which can be represented as follows, including the agent's available resources:

$$Give(Sheep, all) \wedge Receive(Clay, all) \wedge Resources(clay = 0, ore = 0, sheep = 4,$$
$$wheat = 1, wood = 0) \wedge Buildups(roads = 2, settlements = 0, cities = 0).$$

From this illustrative example, y_i=sheep and $\mathbf{x}_i = \{0, 0, 4, 1, 0, 2, 0, 0, 1, 0, 0, 0, 0\}$ based on features 1–13 in Table 1. Although this representation may look simple at first sight, it has support for $8^8 \times 2^5 \times 5 = 2.6$ billion possible (and unique) negotiation events. Notice that not all of them are valid or legal at every point in time in the game. Choosing the most human-like (in our case) trading negotiation can be seen as a *ranking task*, where we focus on computing a score representing the importance of each trading negotiation (similar to the one above) available for choosing the best choice, i.e. the most human-like as seen in the data. In this way, the quality of our learning agents will depend on the quality of the examples provided.

To rank such trading negotiation alternatives, we train a set of statistical classifiers based on the feature set described in Table 1. Our set of features include the resources available (features 1–5), the build-ups (features 6–8) with a default minimum of 0 and maximum value of 7, the receivable resources in binary form to reduce data sparsity (features 9–13), and the givable resource considered as the class prediction (feature 14).

It has to be noted that the feature set listed in Table 1 was chosen for giving the best results from a pool of feature sets. Other feature sets that we explored include smaller domains (only binary features), larger domains (non-binary features), smaller and larger sets of features, and multiple givables, among others.

4   Trading Agents

We cast trading in interactive board games as a classification task, where we compared the following statistical classifiers with the aim of finding the best

Table 1. Feature set for learning trading negotiations from examples.

No.	Feature	Domain	Description
1	hasClay	{0...7}	Number of clay resources available
2	hasOre	{0...7}	Number of ore resources available
3	hasSheep	{0...7}	Number of sheep resources available
4	hasWheat	{0...7}	Number of wheat resources available
5	hasWood	{0...7}	Number of wood resources available
6	hasRoads	{0...7}	Number of roads built so far
7	hasSettlements	{0...7}	Number of settlements built so far
8	hasCities	{0...7}	Number of cities built so far
9	recClay	Binary	1 if clay resources will be received, 0 otherwise
10	recOre	Binary	1 if ore resources will be received, 0 otherwise
11	recSheep	Binary	1 if sheep resources will be received, 0 otherwise
12	recWheat	Binary	1 if wheat resources will be received, 0 otherwise
13	recWood	Binary	1 if wood resources will be received, 0 otherwise
14	givable	Resource	Givable={Clay, Ore, Sheep, Wheat, Wood}

predictor of human-like trades: Bayesian Networks, Conditional Random Fields, and Random Forests.

4.1   Learning to Trade with Bayesian Networks

A Bayesian network represents a joint probability distribution based on a directed acyclic graph, where each node is associated with a probability function and connections represent dependencies. The joint probability distribution over the set of random variables $\mathbf{x} = \{x_1, ..., x_n\}$ is defined by

$$P(\mathbf{x}) = \prod_{i=1}^{n} P(x_i|pa(x_i)), \tag{1}$$

where $pa(.)$ denotes the set of parent random variables, and every variable is associated with a conditional probability distribution $P(x_i|pa(x_i))$. Two main tasks are involved in the creation of our Bayesian network: parameter learning and structure learning. Parameter learning involves the estimation of conditional probability distributions (discrete in our case) from data $D = \{\mathbf{x}_i, y_i\}$ with feature vectors $\mathbf{x}_i$ and labels y_i, where we used maximum likelihood estimation with smoothing. Structure learning involves inducing the dependencies of random variables based on the K2 algorithm, see [5] for details.

Once the Bayes net has been trained, we use the junction tree algorithm [6] for probabilistic inference in order to compute the probabilities of trades. The most probable human-like trade is selected according to

$$y^* = \arg\max_{y \in Y} P(y|evidence(y)), \tag{2}$$

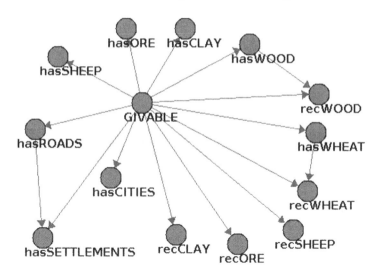

Fig. 2. Bayesian network for predicting human-like trades in board games, where nodes correspond to random variables (contextual information) and arrows represent their dependencies.

where the contextual information of givable y is defined by $evidence(t) = \{f_1 = val_1, ..., f_n = val_n\}$ with features f_i shown in Fig. 2.

4.2 Learning to Trade with Conditional Random Fields

Here, we cast trading in interactive board games as a sequence labeling task, in which a sequence of game environment inputs is labeled with appropriate givable resources to support trades. The task is therefore to find a mapping between (observed) features—including available resources, build-ups, and receivables as shown in Table 1—and a (hidden) sequence of givables.

We use the linear-chain Conditional Random Field (CRF) model for predicting human-like trades in the game of Settlers of Catan, see Fig. 3. This model defines the posterior probability distribution of labels (givables in our case) $\mathbf{y}=\{y_1,\ldots,y_{|\mathbf{y}|}\}$ given features $\mathbf{x}=\{x_1,\ldots,x_{|\mathbf{x}|}\}$, as

$$P(\mathbf{y}|\mathbf{x}) = \frac{1}{Z(\mathbf{x})} \prod_{t=1}^{T} \exp\left\{\sum_{k=1}^{K} \theta_k \Phi_k(y_t, y_{t-1}, \mathbf{x}_t)\right\}, \tag{3}$$

where $Z(\mathbf{x})$ is a normalization factor over all available vectors of contextual information $\mathbf{x}$ such that the sum of all labelings is one. The parameters θ_k are weights associated with feature functions $\Phi_k(.)$, which are real values describing the label state y at time t based on the previous label state y_{t-1} and features $\mathbf{x}_t$. For example: from Eq. 3, Φ_k might have the value $\Phi_k = 1.0$ for the transition from *"Give(Sheep)"* to *"Give(Clay)"*, and 0.0 elsewhere. The parameters θ_k are

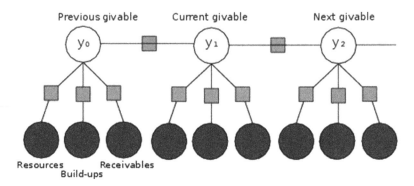

Fig. 3. Conditional Random Field (CRF) for predicting human-like trades in board games, where empty nodes correspond to the labeled sequence (givables: Clay, Ore, Sheep, Wheat, Wood), shaded nodes to game features, and squares represent relationships between labels and features.

set to maximize the conditional likelihood of sequences of givables in the training data set. They are estimated using the gradient descent algorithm.

After training, labels can be predicted for new sequences of observations. The most likely trading offer is expressed as:

$$y^* = \arg\max_y Pr(y|\mathbf{x}), \tag{4}$$

which is computed using the Viterbi and A* search algorithms, see [19] for details.

4.3 Learning to Trade with Random Forests

This agent is trained using an ensemble of trees as shown in Fig. 4, which are used to vote for the class prediction at test time [3]. A random forest is an ensemble learning method that constructs a set of random decision trees at training time, and uses them to generate the most popular class. Random forests are attractive due to their ability to offer better generalization (i.e. less overfitting to data) than other techniques such as decision trees [17]. We compute the probability distribution of a human-like trade as:

$$P(givable|evidence) = \frac{1}{Z} \prod_{b \in B} P_b(givable|evidence), \tag{5}$$

where *givable* refers to the class prediction, *evidence* refers to observed features 1–13, $P_b(.|.)$ is the posterior distribution of the bth tree, and Z is a normalization constant—see [7] for further details. Assuming that Y is a set of givables at a particular point in time in the game, extracting the most human-like trading offer (givable y^*) and collected evidence (context of the game), is defined as

$$y^* = \arg\max_{y \in Y} Pr(y|evidence). \tag{6}$$

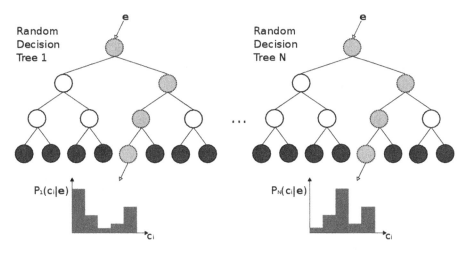

Fig. 4. Random forest for predicting human-like trades, where non-leaf nodes represent contextual information (**e**) and leaf nodes represent the class prediction (givables)—adapted from [4].

5   Experiments and Results

We first describe the evaluation metrics used to assess the performance of the statistical classifiers described above. An offline evaluation is then described to report performance on held-out data, and finally an online evaluation is described to assess performance while playing the game of Settlers of Catan using a benchmark framework.

5.1   Evaluation Metrics

The evaluation metrics that we use to assess the predictive power of human-like trading include classification accuracy and precision-recall. The former is computed as

$$\text{Accuracy} = \frac{t_p + t_n}{t_p + t_n + f_p + f_n};$$ (7)

and the latter as

$$\text{F-measure} = \frac{2 \times precision \times recall}{precision + recall},$$ (8)

where

$$precision = \frac{t_p}{t_p + f_p},$$ (9)

$$recall = \frac{t_p}{t_p + f_n},$$ (10)

and t_p=true positives, t_n=true negatives, f_p=false positives, and f_n=false negatives.

Table 2. Precision-Recall results of human trading negotiations in Settlers of Catan.

Classifier	Accuracy (%)	Precision	Recall	F-Measure
Majority Baseline	23.43	0.055	0.234	0.089
Conditional Random Field	62.08	0.623	0.623	0.623
Bayesian Network	63.90	0.640	0.639	0.635
Random Forest	**65.72**	**0.657**	**0.657**	**0.656**

In addition, to assess the agent's performance while playing the game we consider the following game-related metrics (in terms of averages): percentage of winning rate, victory points, offers made, successful offers, and pieces built.

5.2 Offline Evaluation

Table 2 shows the classification results of our statistical classifiers using the features listed in Table 1 trained on the data described in Sect. 3. This evaluation is based on 10-fold cross validations. The first thing that can be noted is that predicting human trades is a difficult task because our best classifier, the Random Forest, achieves a classification accuracy of 65.7 %. A second thing to notice is that our statistical classifiers (Bayes Nets, CRFs, and Random Forest) substantially outperform a majority baseline. These results motivate future work on learning agents with improved performance.

5.3 Online Evaluation

We also evaluated the statistical classifiers described in Sect. 4 using the JSettlers interface [29], where we use a baseline rule-based negotiator[1] as the opponent. Their integration is illustrated in Fig. 5. We refer to this evaluation as 'online' because the agents were used in the actual game to rank realistic trading negotiations. Although this evaluation had the GUI turned off, in games with human players the GUI would simply have been turned on. This means that all games were run using four automated agents: one statistical vs. three rule-based. We evaluate each classifier with 10,000 games in order to obtain significant comparisons due to the randomness exhibited in the game. Such a number of games has shown to produce meaningful comparisons [16].

Table 3 shows the results of this online evaluation. It can be observed that the baseline agents obtain a winning rate of 25 % because four players of the same kind play against each other. It can also be observed that not all agents using the trained classifiers outperform the rule-based agents resulting in higher winning rates, more victory points, and more pieces built, but not necessarily more offers.

---

[1] The baseline agent (referred to as 'rule-based') included the following parameters in all agents, see [16] for further details: TRY_N_BEST_BUILD_PLANS:0, FAVOUR_DEV_CARDS:-5.

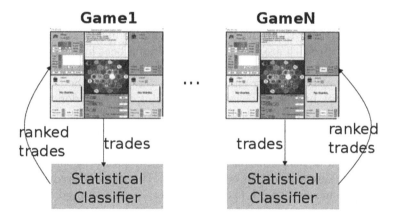

Fig. 5. High-level integration of statistical classifiers with the JSettlers interface [29]

We can also note that the best results are obtained by the Random Forest classifier. These results extend the previous section and reveal that classifiers with higher accuracy also achieved higher win rates.

We also note that the training data was not collected from particularly experienced or expert players of the game—and so we can expect that training on a corpus of expert-player trades would have achieved still better performance.

Table 3. Game results when comparing statistical classifiers against three rule-based traders, i.e. four players in each game—each line shows average results over 10,000 games.

Comparison between trained statistical trader vs opponent	Wining rate (%)	Victory points	Offers made	Successful offers	Pieces built
Rule-based vs rule-based	25.00	6.40	**147.57**	**137.95**	8.33
Conditional Random Field vs rule-based	23.31	6.20	141.39	131.89	7.96
Bayesian Network vs rule-based	24.20	6.20	141.59	131.72	7.98
Random Forest vs rule-based	**27.62**	**6.54**	145.61	135.84	**8.50**

6   Related Work

Machine learning techniques for strategic board games have received little attention so far. Notable exceptions have applied Reinforcement Learning to board games:

– [28] proposes reinforcement learning with multilayer neural networks for training an agent to play the game of Backgammon. He finds that trained agents with such an approach are able to match and even beat human performance.

– [23] proposes hierarchical reinforcement learning for automatic decision making on object-placing and trading actions in the game of Settlers of Catan. He incorporates built-in knowledge for learning the behaviors of the game quicker, and finds that the combination of learned and built-in knowledge is able to beat human players.

– More recently, [11] used reinforcement learning in non-cooperative dialogue, and focused on a small 2-player trading problem with 3 resource types, and without using any real human dialogue data. This work showed that explicit manipulation moves (e.g. "I really need sheep") can be used to win when playing against adversaries who are gullible (i.e. they believe such statements) but also against adversaries who can detect manipulation and can punish the player for being manipulative [10]. Strategies for beating opponent models in trading have also been explored recently [12].

Some supervised learning techniques have also been applied to board games such as decision trees [13], preference learning [25], and deep neural networks [21]. Since statistical inference has been ignored in previous work, with some exceptions such as [21], we argue that it can play an important role in training strategic agents with human-like behavior. In addition and to the best of our knowledge, the Random Forest classifier has not been applied to strategic board games before, and our results report that it represents a state-of-the-art method for learning trading negotiations.

Other forms of machine learning that can be explored include not only direct but also inverse reinforcement learning to learn from trial and error, semi-supervised learning to learn from labeled and unlabeled data, unsupervised learning to learn from unlabeled data, multi-agent systems to learn behaviors considering the strategies of opponents, transfer learning so that agents do not have to be trained from scratch, active learning to learn to ask what to do in uncertain situations while playing the game, among others—see [8,13,24] for an overview. Another direction to explore in strategic games includes a combination of planning and learning, which has shown more promising results than any of them in isolation [21]. A further direction includes the joint optimization of game behavior and the corresponding verbalizations [9,20].

Other related work has been carried out in the context of automated non-cooperative dialogue systems, where an agent may act to satisfy its own goals rather than those of other participants [14,15,18]. The game-theoretic underpinnings of non-cooperative behavior have also been investigated [2]. Such automated agents are of interest when trying to persuade, argue, or debate, or in the area of believable characters in video games and educational simulations [15,26]. Another arena in which non-cooperative dialogue behavior has been investigated is in negotiation [30], where hiding information (and even outright lying) can be advantageous.

7   Conclusions and Future Work

This paper presents an approach for learning to trade in strategic board games based on training examples collected from human players. To do that, we train three statistical classifiers (Bayesian Network, Conditional Random Field, and Random Forest) in a supervised manner, and then apply statistical inference in order to compute probabilistic scores for each trading negotiation available at a particular point in the game. Those scores are used to rank the available trading negotiations, where the top choice (i.e. the most human-like) is used in the game. In an offline evaluation, the statistical classifiers show that there is still substantial room for improvement, where the best classification score (65.7 %) was obtained by the Random Forest. In an online evaluation, the same ranking of classifiers was observed—Random Forest obtaining the best results. Our results are encouraging for training classifiers with improved performance in order to incorporate highly strategic behavior in trading.

Future directions include:

- training statistical agents that take into account richer contextual information such as features from other players, and training them to play multiple games;
- exploring other forms of machine learning, as mentioned above; and
- evaluating trained agents against human players.

Acknowledgments. Funding from the European Research Council (ERC) project "STAC: Strategic Conversation" no. 269427 is gratefully acknowledged. See http://www.irit.fr/STAC/. We would like to thank the following members of the STAC project for helpful discussions: Markus Guhe, Eric Kow, Mihai Dobre, Ioannis Efstathiou, Verena Rieser, Alex Lascarides, and Nicholas Asher.

References

1. Afantenos, S., Asher, N., Benamara, F., Cadilhac, A., Dégremont, C., Denis, P., Guhe, M., Keizer, S., Lascarides, A., Lemon, O., Muller, P., Paul, S., Rieser, V., Vieu, L.: Developing a corpus of strategic conversation in the Settlers of Catan. In: Workshop on the Semantics and Pragmatics of Dialogue SeineDial, Paris, France (2012). https://hal.inria.fr/hal-00750618
2. Asher, N., Lascarides, A.: Commitments, beliefs and intentions in dialogue. In: Proceedings of SemDial, pp. 35–42 (2008)
3. Breiman, L.: Random forests. Mach. Learn. **45**(1), 5–32 (2001)
4. Choi, J., Park, J., Park, H., Park, J.I.: iHand: an interactive bare-hand-based augmented reality interface on commercial mobile phones. Opt. Eng. **52**(2), 027206 (2013)
5. Cooper, G., Herskovits, E.: A Bayesian method for the induction of probabilistic networks from data. Mach. Learn. **9**(4), 309–347 (1992)
6. Cozman, F.G.: Generalizing variable elimination in Bayesian networks. In: Workshop on Probabilistic Reasoning in Artificial Intelligence, pp. 27–32 (2000)

7. Criminisi, A., Shotton, J., Konukoglu, E.: Decision forests: a unified framework for classification, regression, density estimation, manifold learning and semi-supervised learning. Found. Trends Comput. Graph. Vis. **7**(2–3), 81–227 (2012)
8. Cuayáhuitl, H., van Otterlo, M., Dethlefs, N., Frommberger, L.: Machine learning for interactive systems and robots: a brief introduction. In: 2nd Workshop on Machine Learning for Interactive Systems (MLIS), pp. 19–28. ACM (2013)
9. Dethlefs, N., Cuayáhuitl, H.: Hierarchical reinforcement learning for situated natural language generation. Nat. Lang. Eng. **21**(3), 391–435 (2015)
10. Efstathiou, I., Lemon, O.: Learning to manage risk in non-cooperative dialogues. In: Proceedings of SEMDIAL (2014)
11. Efstathiou, I., Lemon, O.: Learning non-cooperative dialogue behaviours. In: SIG-DIAL (2014)
12. Efstathiou, I., Lemon, O.: Learning non-cooperative dialogue policies to beat opponent models: 'the good, the bad and the ugly'. In: Proceedings of SEMDIAL (2015)
13. Fürnkranz, J.: Machine learning in games: a survey. In: Machines that Learn to Play Games, Chapter 2, pp. 11–59. Nova Science Publishers (2000)
14. Georgila, K., Nelson, C., Traum, D.: Single-agent vs. multi-agent techniques for concurrent reinforcement learning of negotiation dialogue policies. In: Proceedings of the 52nd Annual Meeting of the Association for Computational Linguistics, Baltimore, USA, pp. 500–510, September 2014
15. Georgila, K., Traum, D.: Reinforcement learning of argumentation dialogue policies in negotiation. In: Proceedings of INTERSPEECH (2011)
16. Guhe, M., Lascarides, A.: Game strategies for the Settlers of Catan. In: 2014 IEEE Conference on Computational Intelligence and Games, CIG 2014, Dortmund, Germany, 26–29 August 2014, pp. 1–8 (2014)
17. Hastie, T., Tibshirani, R., Friedman, J.: The Elements of Statistical Learning: Data Mining, Inference and Prediction, 2nd edn. Springer, New York (2009)
18. Hiraoka, T., Georgila, K., Nouri, E., Traum, D., Nakamura, S.: Reinforcement learning in multi-party trading dialog. In: Proceedings of the SIGDIAL 2015 Conference, Prague, Czech Republic, pp. 32–41, September 2015
19. Kudo, T.: CRF++: Yet another CRF toolkit (2005). http://crfpp.sourceforge.net
20. Lemon, O.: Adaptive natural language generation in dialogue using reinforcement learning. In: Proceedings of SEMDIAL (2008)
21. Maddison, C.J., Huang, A., Sutskever, I., Silver, D.: Move evaluation in go using deep convolutional neural networks. CoRR abs/1412.6564 (2014)
22. McFarlin, M.: 10 great board games for traders. Futures Magazine (2013). http://www.futuresmag.com/2013/10/02/10-great-board-games-for-traders
23. Pfeiffer, M.: Reinforcement learning of strategies for Settlers of Catan. In: International Conference on Computer Games: Artificial Intelligence, Design and Education (2004)
24. Pietquin, O., Lopez, M.: Machine learning for interactive systems: challenges and future trends. In: Proceedings of the Workshop Affect, Compagnon Artificiel (WACAI) (2014)
25. Runarsson, T.P., Lucas, S.M.: Preference learning for move prediction and evaluation function approximation in Othello. IEEE Trans. Comput. Intell. AI Games **6**(3), 300–313 (2014)
26. Shim, J., Arkin, R.: A taxonomy of robot deception and its benefits in HRI. In: Proceedings of IEEE Systems, Man and Cybernetics Conference (2013)
27. Szita, I., Chaslot, G., Spronck, P.: Monte-Carlo tree search in Settlers of Catan. In: van den Herik, H.J., Spronck, P. (eds.) ACG 2009. LNCS, vol. 6048, pp. 21–32. Springer, Heidelberg (2010)

28. Tesauro, G.: Temporal difference learning and TD-Gammon. Commun. ACM **38**(3), 58–68 (1995)
29. Thomas, R., Hammond, K.J.: Java settlers: a research environment for studying multi-agent negotiation. In: Intelligent User Interfaces (IUI), p. 240 (2002)
30. Traum, D.: Extended abstract: computational models of non-cooperative dialogue. In: Proceedings of SIGdial Workshop on Discourse and Dialogue (2008)

Argumentative AI Director Using Defeasible Logic Programming

Ramiro A. Agis, Andrea Cohen[(✉)], and Diego C. Martínez

Artificial Intelligence Research and Development Laboratory (LIDIA),
Department of Computer Science and Engineering (DCIC),
Universidad Nacional del Sur (UNS),
Consejo Nacional de Investigaciones Científicas y Técnicas (CONICET),
Bahía Blanca, Argentina
`ramiro.agis@live.com`, {`ac,dcm`}`@cs.uns.edu.ar`

Abstract. In this work we present a novel implementation of an AI Director that uses argumentation techniques to decide dynamic adaptations in the level generation of a roguelike game called *HermitArg*. The architecture of the game introduces *smart items* with defeasible information to be analyzed in a dialectical process.

1 Introduction

The use of adaptive elements in a game level is a novel technique intended to improve the user experience. This can be achieved by including an overseer agent, usually called *AI Director* [2,6,14], whose goal is to keep the user interested from the beginning to the end while providing a fair and balanced level of challenge. The Director agent monitors player activity data, and then it adjusts some game parameters to provide a different experience that fits perfectly with what the player is capable of. For instance, in the game *Left 4 Dead* (by *Valve Corporation*), the AI Director decides where weapons, ammo and useful items are spawned. It also decides when and where a desperate horde of angry zombies (an *infected rush*) starts. Adaptive games also contribute to *player retention* since the challenge of the game is adjusted as the user plays. Player retention is the ability to keep the user interested in the game for a long period of time. It is an important feature of game development today, especially in long-life games, like massively-multiplayer online role playing games (MMORPGs). The reasons why players decide to return to a game are different for every player and game genre, but undoubtedly it implies that the challenge imposed by the ludic component of the game is still *active and attractive*. There is now a modest, yet growing research activity about player retention through several artificial intelligent formalisms, but none of them integrating *argumentation* techniques.

Defeasible Argumentation is a form of reasoning where arguments for and against a proposition are produced and evaluated to verify the acceptability of that proposition. An argument is a tentative piece of reasoning supporting a claim. The main idea in argumentation systems is that any proposition will be

© Springer International Publishing Switzerland 2016
T. Cazenave et al. (Eds.): CGW 2015/GIGA 2015, CCIS 614, pp. 96–111, 2016.
DOI: 10.1007/978-3-319-39402-2_8

accepted as true if there exists an argument supporting it, and this argument is acceptable according to an analysis considering all its counter-arguments. Argumentation is suitable to deal with incomplete and contradictory information in dynamic domains. In particular, *Defeasible Logic Programming (DeLP)* [5], is a logic programming paradigm based on argumentation that allows for the representation of strict and defeasible knowledge to construct arguments providing reasons to defeasible claims. It is a concrete argumentation formalism with a solid implementation (see *e. g.*, [3,15]).

In this work we present an implementation of an Argumentative AI Director that uses argumentation to decide dynamic adaptations of a roguelike game called *HermitArg*. This Director uses Defeasible Logic Programming to represent information about the state of the game, in order to *argue* about the complexity of the next level. We also define an architecture to customize the game by providing a level item with a piece of defeasible knowledge about the adaptability of such an item. Hence, the Director will evaluate the adaptation of this *smart* item by contrasting this defeasible information with the whole knowledge base.

This work is organized as follows. We first review DeLP, the argumentation formalism used by the AI Director proposed in this paper. In Sect. 3 we present HermitArg and explain the general behavior of its AI Director. Section 4 explains how the AI Director uses the argumentation formalism presented in Sect. 2 as part of its decision making process. In Sect. 5 we present the design and implementation of our Argumentative AI Director. Section 6 discusses related work. In Sect. 7 we present conclusions and comment on future lines of work.

2   Defeasible Logic Programming (DeLP)

In the last decades, argumentation has evolved as an attractive paradigm for conceptualizing common-sense reasoning [12]. As a result, several approaches were proposed to model argumentation on an abstract basis [4], using classical logics [1], or using logic programming [5], among others. Next, we include a short explanation of Defeasible Logic Programming (DeLP), the structured argumentation system of [5], which will be used as a knowledge representation and reasoning formalism for the *Argumentative AI Director* proposed in this paper.

Similarly to Logic Programming, DeLP represents information using facts and (strict) rules. In addition, DeLP has the declarative capability of representing weak information in the form of defeasible rules, and a defeasible argumentation inference mechanism for warranting the entailed conclusions. Thus, knowledge in DeLP will be represented through *facts*, *strict rules*, and *defeasible rules*, defined as follows. *Facts* are ground literals representing atomic information, or the negation of atomic information using *strong negation* "$\sim$" (*e. g.*, $chicken(tina)$ or $\sim dog(tina)$). *Strict Rules* are denoted $L_0 \leftarrow L_1, \ldots, L_n$, where L_0 is a ground literal and $\{L_i\}_{i>0}$ is a finite set of ground literals (*e. g.*, $bird(tina) \leftarrow chicken(tina)$). *Defeasible Rules* are denoted $L_0 \prec L_1, \ldots, L_n$, where L_0 is a ground literal and $\{L_i\}_{i>0}$ is a set of ground literals (*e. g.*, $flies(tina) \prec bird(tina)$ or $\sim flies(tina) \prec chicken(tina)$).

Syntactically, the only difference between strict and defeasible rules is the kind of arrow they use (respectively, " $\leftarrow$ " and " $\prec$ "). However, while strict rules represent non-defeasible information, defeasible rules represent tentative information that may be used if nothing could be posed against them. That is, a defeasible rule "$Head \prec Body$" expresses that "*reasons to believe in the antecedent Body give reasons to believe in the consequent Head*".

A *Defeasible Logic Program* (*DeLP program*, for short) $\mathcal{P}$ is a set of facts, strict rules and defeasible rules. When required, $\mathcal{P}$ will be noted as (Π, Δ), distinguishing the subset Π of facts and strict rules, and the subset Δ of defeasible rules. Strict and defeasible rules are ground, however, following the usual convention [9], examples will use "schematic rules" with variables denoted with an uppercase letter.

Strong negation could appear in the head of program rules, and can be used to represent contradictory knowledge. Therefore, from a program $\mathcal{P}$ complementary literals could be derived, however, the set Π used to represent non-defeasible information must be non-contradictory (*i. e.*, no pair of complementary literals could be derived from Π). For the treatment of contradictory knowledge, DeLP incorporates a defeasible argumentation formalism. This formalism allows for the identification of the pieces of knowledge that are in contradiction, and a *dialectical process* is used for deciding which information prevails as warranted. The dialectical process involves the construction and evaluation of *arguments* that either support or interfere with the query under analysis.

Briefly, an *argument* for a literal h, denoted $\langle \mathcal{A}, h \rangle$, is a minimal set of defeasible rules $\mathcal{A} \subseteq \Delta$ such that $\mathcal{A} \cup \Pi$ is non-contradictory and there is a derivation for h from $\mathcal{A} \cup \Pi$. Intuitively, a literal h will be *warranted* from $\mathcal{P}$ if there exists an undefeated argument $\mathcal{A}$ supporting h. To establish if $\langle \mathcal{A}, h \rangle$ is an undefeated argument, *counter-arguments* that could be defeaters for $\langle \mathcal{A}, h \rangle$ are considered. Then, a counter-argument will be a *defeater* if it is preferred to the attacked argument $\langle \mathcal{A}, h \rangle$, according to an argument comparison criterion.

In this work we will adopt *generalized specificity* [5], a comparison criterion that prefers more precise or more direct arguments. When comparing two arguments $\langle \mathcal{A}_1, h_1 \rangle$ and $\langle \mathcal{A}_2, h_2 \rangle$ the comparison criterion performs an analysis of their activation. Briefly, a set of literals S activates an argument $\langle \mathcal{A}, h \rangle$ if the literal h can be derived from the set $\Pi_r \cup S \cup \mathcal{A}$, where Π_r is the set of program strict rules. Then, $\langle \mathcal{A}_1, h_1 \rangle$ is more specific than $\langle \mathcal{A}_2, h_2 \rangle$ if every set of literals H that non-trivially activates $\langle \mathcal{A}_1, h_1 \rangle$ also activates $\langle \mathcal{A}_2, h_2 \rangle$, and there exists a set of literals H' that non-trivially activates $\langle \mathcal{A}_2, h_2 \rangle$ but H' does not activate $\langle \mathcal{A}_1, h_1 \rangle$. Notwithstanding this, as explained by the authors in [5], DeLP's comparison criterion is modular and thus, it could be easily changed.

Since defeaters are arguments, there may be defeaters for them, defeaters for those defeaters, and so on. Thus, exhaustive sequences of arguments called *argumentation lines* are constructed, where each argument defeats its predecessor in the line, and no (sub)arguments can be reintroduced in the same line. All argumentation lines starting with $\langle \mathcal{A}, h \rangle$ are then grouped together into a *dialectical tree* rooted in $\langle \mathcal{A}, h \rangle$. Then, starting from the leaves up to the root,

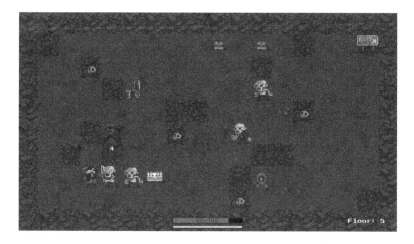

Fig. 1. General setup of a level in HermitArg.

each node (argument) in the tree will be marked as *defeated* (D) or *undefeated* (U) as follows. Leaves represent arguments for which no defeaters can be added to the corresponding argumentation line and thus, they will be marked U. If an inner node has at least one child marked U, then it is marked D; otherwise, it will be marked U (for a detailed explanation of DeLP's dialectical process see [5]). Finally, a literal h will be *warranted* if there exists an argument $\langle \mathcal{A}, h \rangle$ that is marked U in the dialectical tree rooted in it.

Given a DeLP program $\mathcal{P}$, it is possible to make queries in order to find out whether the literals represented by those queries are warranted or not. Consequently, a query Q could have one of the following answers: *YES*, if the literal Q is warranted from $\mathcal{P}$; *NO*, if the literal $\overline{Q}$ (*i.e.*, the complement of Q with respect to the strong negation "$\sim$") is warranted from $\mathcal{P}$; *UNDECIDED*, if neither Q nor $\overline{Q}$ is warranted from $\mathcal{P}$; and *UNKNOWN*, if Q does not belong to the signature of the program $\mathcal{P}$.

3   HermitArg: An Argumentative AI Director Case Study

HermitArg is a turn-based roguelike game developed in *Unity 2D* in which the player controls a hermit that has to escape from a cave. As the game goes by, the player will venture deeper into the cave, which is designed as a labyrinth with several floors. Each level in the game corresponds to a different floor of the cave, where the player has to fight monsters, break walls and avoid obstacles in order to reach the exit. Along the way, he can pick up food and booze to recover health and open chests containing power-ups to increase his chances of defeating the enemies. Figure 1 shows a screenshot of HermitArg, distinguishing the hermit and its corresponding health, walls, different kinds of enemies and chests.

HermitArg implements an *AI Director* whose task is to procedurally generate every new level depending on the player's status and performance in the previous one. As mentioned before, the goal of an AI Director is to keep the user interested from the beginning to the end of the game, while providing a fair and balanced level of challenge. Thus, using player activity data, the AI Director will automatically and dynamically change difficulty, scenarios, and behavior in the game. For instance, if the hermit is about to die or badly hurt, the AI Director will include more food and reduce the amount of enemies (with respect to the standard values) in the next level. On the other hand, if the player managed to kill all enemies and is fully or almost fully recovered, it means that he had a really good performance in that level and, consequently, the next level will be harder than usual.

Over the past 7 years, since *Valve* released *Left 4 Dead*, the notion of AI Director has not only gained traction in the video game industry but also established itself as one of the best practices for building a high quality gameplay. As a result, this concept was also implemented in other games such as *Saints Row IV* (Dev. *Volition*, Pub. *Deep Silver*) and *Darkspore* (Dev. *Maxis*, Pub. *Electronic Arts*). However, none of the existing games so far have incorporated an argumentative mechanism to aid the AI Director's decision making process.

The singularity of HermitArg's AI Director is that it uses DeLP as a knowledge representation and reasoning formalism. Briefly, in order to build a new level, it creates a knowledge base composed of the game's decision mechanics and relevant state in the form of a DeLP program (*i. e.*, a set of facts, strict rules and defeasible rules). Then, the AI Director will perform different queries on the knowledge base in order to decide how to generate the next level.

The alternative results of the queries performed by the AI Director may lead to different courses of action, requiring the modification of different parameters in the game. Among these parameters, we can distinguish the amount of food and booze dropped on the floor, as well as the quantity of enemies to be faced. In particular, these parameters are initialized with standard values, which may be adapted as a result of the Director's decision making process. In the case of food & booze, the standard count equals the level number minus 2 (or zero, in the first two levels); on the other hand, the standard value for enemy count coincides with the level number. Then, once the final values are obtained, the Director proceeds to the generation of the new level following these parameters.

4   DeLP Entities and AI Director's Reasoning Process

Every element of the game considered during the AI Directors's decision making process is referred to as a *DeLP Entity*. A DeLP Entity is composed of four attributes: a set of possible facts, a set of strict rules, a set of defeasible rules and a set of associated queries. These rules and facts will be added to the Director's knowledge base, which will then be queried by the different DeLP Entities before the creation of a new level. On the one hand, the resulting knowledge base will be such that it contains every rule (strict and defeasible) from every entity in the game. On the other hand, each entity will be responsible for determining the conditions under which each of its own facts will be added to the knowledge

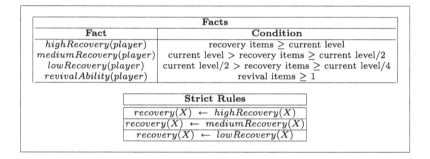

Facts	
Fact	**Condition**
$highRecovery(player)$	recovery items $\geq$ current level
$mediumRecovery(player)$	current level > recovery items $\geq$ current level/2
$lowRecovery(player)$	current level/2 > recovery items $\geq$ current level/4
$revivalAbility(player)$	revival items ≥ 1

Strict Rules
$recovery(X) \leftarrow highRecovery(X)$
$recovery(X) \leftarrow mediumRecovery(X)$
$recovery(X) \leftarrow lowRecovery(X)$

Fig. 2. Description of DeLP Entity *Inventory*.

base, as well as the conditions for making each query and specifying the actions to be performed depending on their results.

For instance, let us consider two of the main DeLP Entities in HermitArg. The first one, called *Inventory*, models information about the different power-ups and items the player has, such as recovery or revival items. The second entity, called *Player*, models different aspects of the hermit's status and performance. Figures 2 and 3 illustrate a simplified version of these DeLP Entities in HermitArg.

It can be noted that the *Inventory* entity has no associated queries. Hence, the information modelled by its strict rules and facts will be used in the resolution of queries from other entities. The fact *"revivalAbility(player)"* will be added into the Director's knowledge base if and only if the player has any revival items in the inventory. On the other hand, as shown in Fig. 2, the conditions associated with the facts *"highRecovery(player)"*, *"mediumRecovery(player)"* and *"lowRecovery(player)"* are exclusionary. Thus, the Director's knowledge base will only include one of these, modelling the highest recovery capability of the player. Finally, the strict rules express that if the player has reached the minimum threshold of recovery items, then it has the ability to recover.

Differently from the inventory, all attributes of the DeLP Entity *Player* are specified. The conditions associated with the facts *"fullHealth(player)"*, *"highHealth(player)"*, *"mediumHealth(player)"*, *"lowHealth(player)"* and *"veryLowHealth(player)"* are exclusionary. Thus, the Director's knowledge base will only include one of these, modelling the current health level of the player. On the other hand, the fact *"experienced(player)"* represents that the player is at an advanced level of the game. Thus, this fact will only be added to the Director's knowledge base when the current level is higher than 10. The strict rules express that if the player's health is over 50 %, then it has no risk of dying. The first six defeasible rules give defeasible reasons to believe that the player's life is at risk (respectively, not at risk) when its health level drops below 50 %. In particular, when having *low health*, the player may be saved from dying if it will revive or has, at least, the amount of recovery items required to reach a low recovery level. In contrast, when its health is *very low*, it may only be saved by reviving, or by having the amount of recovery items required to reach a high

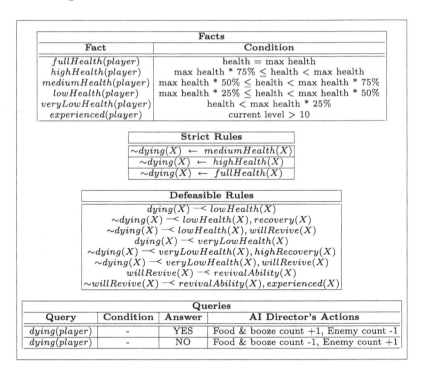

Facts	
Fact	**Condition**
$fullHealth(player)$	health = max health
$highHealth(player)$	max health * 75% $\leq$ health < max health
$mediumHealth(player)$	max health * 50% $\leq$ health < max health * 75%
$lowHealth(player)$	max health * 25% $\leq$ health < max health * 50%
$veryLowHealth(player)$	health < max health * 25%
$experienced(player)$	current level > 10

Strict Rules
$\sim dying(X) \leftarrow mediumHealth(X)$
$\sim dying(X) \leftarrow highHealth(X)$
$\sim dying(X) \leftarrow fullHealth(X)$

Defeasible Rules
$dying(X) \prec lowHealth(X)$
$\sim dying(X) \prec lowHealth(X), recovery(X)$
$\sim dying(X) \prec lowHealth(X), willRevive(X)$
$dying(X) \prec veryLowHealth(X)$
$\sim dying(X) \prec veryLowHealth(X), highRecovery(X)$
$\sim dying(X) \prec veryLowHealth(X), willRevive(X)$
$willRevive(X) \prec revivalAbility(X)$
$\sim willRevive(X) \prec revivalAbility(X), experienced(X)$

Queries			
Query	**Condition**	**Answer**	**AI Director's Actions**
$dying(player)$	-	YES	Food & booze count +1, Enemy count -1
$dying(player)$	-	NO	Food & booze count -1, Enemy count +1

Fig. 3. Description of DeLP Entity *Player*.

recovery level. Then, the last two defeasible rules respectively provide reasons for and against the player's revival: the former expresses that the player will revive if it has the ability to do so; whereas the latter expresses that, even though the player has the ability to revive, it will not revive when being at an advanced state of the game. Finally, the query associated with the *Player* entity allows the AI Director to increase or decrease the difficulty of the next level, depending on whether the player was dying at the end of the previous level or not.

It could be noted that the query of the *Player* entity has no associated condition. This is because the AI Director will be interested in knowing whether the player was dying or not before creating *every* new level. In contrast, for instance, let us suppose that the *Player* entity also specifies the query *"tooStrong(player)"* related to the player's high killing performance in the last level, where a positive answer leads the Director to significantly increase the amount of enemies while dramatically reducing the amount of food & booze in the following level. Taking this into account, the AI Director may only be interested in asking that query under specific circumstances, such as if the player completed the previous level without being close to dying. Otherwise, if that query is asked even though the player was close to death, and the answer is positive, then it would most likely mean that the player will get killed in the next level (as a result of having to face a greater amount of monsters and having less chances of recovery).

The following example presents a concrete scenario of HermitArg, illustrating how the AI Director's knowledge base is obtained.

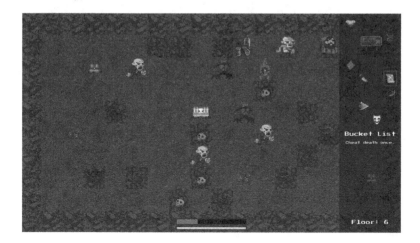

Fig. 4. HermitArg's scenario where the player is about to complete level 6.

Example 1. *Let us consider the scenario depicted in Fig. 4, where the player is about to complete level 6 having a 30 % of health, no recovery items and one revival item (the Bucket List) in its inventory. Since the hermit's health ranges between 25 %–50 % of its total, the DeLP Entity* Player *will add the fact "lowHealth(player)" to the AI Director's knowledge base. Also, since the hermit has no recovery power-ups but has a revival item, the DeLP Entity* Inventory *will add the fact "revivalAbility(player)". Finally, since the current level is 6, the fact "experienced(player)" will not be added to the knowledge base. Thus, the AI Director's knowledge base (state of the game at the end of level 6) is:*

$$\mathcal{P}_6 = \left\{ \begin{array}{l} lowHealth(player) \\ \sim dying(X) \leftarrow mediumHealth(X) \\ \sim dying(X) \leftarrow highHealth(X) \\ \sim dying(X) \leftarrow fullHealth(X) \\ dying(X) \prec lowHealth(X) \\ \sim dying(X) \prec lowHealth(X), recovery(X) \\ \sim dying(X) \prec lowHealth(X), willRevive(X) \\ dying(X) \prec veryLowHealth(X) \\ \sim dying(X) \prec veryLowHealth(X), highRecovery(X) \\ \sim dying(X) \prec veryLowHealth(X), willRevive(X) \\ willRevive(X) \prec revivalAbility(X) \\ \sim willRevive(X) \prec revivalAbility(X), experienced(X) \\ revivalAbility(player) \\ recovery(X) \leftarrow highRecovery(X) \\ recovery(X) \leftarrow mediumRecovery(X) \\ recovery(X) \leftarrow lowRecovery(X) \end{array} \right\}$$

After generating the knowledge base, it is time for the AI Director to ask, when corresponding, the queries associated to the DeLP Entities and perform certain actions depending on their results. The following example illustrates the dialectical process carried out in order to answer a DeLP query.

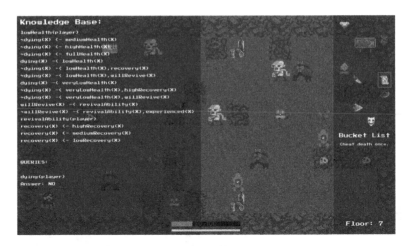

Fig. 5. Screenshot of HermitArg, illustrating the beginning of level 7.

Example 2. *Given the scenario described in Example 1, the AI Director will make the query "dying(player)" before generating level 7. In order to resolve this query, we have to consider all the arguments built from the AI Director's knowledge base $\mathcal{P}_6$ whose conclusion is "dying(player)". In this case, the only argument built from $\mathcal{P}_6$ meeting that criterion is $\langle \mathcal{A}_1, dying(player) \rangle$, with $\mathcal{A}_1 = \{dying(player) \prec lowHealth(player)\}$. Then, we search for possible counter-arguments for $\langle \mathcal{A}_1, dying(player) \rangle$, i.e., arguments whose conclusion is "$\sim$dying(player)". The only argument built from $\mathcal{P}_6$ meeting that criterion is $\langle \mathcal{A}_2, \sim dying(player) \rangle$, with $\mathcal{A}_2 = \{(willRevive(player) \prec revivalAbility(player)), (\sim dying(player) \prec lowHealth(player), willRevive(player))\}$. Now we have to determine whether the attack from $\langle \mathcal{A}_2, \sim dying(player) \rangle$ succeeds, in which case $\langle \mathcal{A}_2, \sim dying(player) \rangle$ will be a defeater of $\langle \mathcal{A}_1, dying(player) \rangle$. As mentioned in Sect. 2, we will adopt generalized specificity as the argument comparison criterion. Therefore, since $\langle \mathcal{A}_2, \sim dying(player) \rangle$ is preferred to $\langle \mathcal{A}_1, dying(player) \rangle$ because of being more precise, $\langle \mathcal{A}_2, \sim dying(player) \rangle$ is a defeater of $\langle \mathcal{A}_1, dying(player) \rangle$. Given that the only counter-argument of $\langle \mathcal{A}_2, \sim dying(player) \rangle$ is $\langle \mathcal{A}_1, dying(player) \rangle$, $\langle \mathcal{A}_2, \sim dying(player) \rangle$ has no defeaters. Hence, the only argumentation line starting with $\langle \mathcal{A}_1, dying(player) \rangle$ is $[\langle \mathcal{A}_1, dying(player) \rangle, \langle \mathcal{A}_2, \sim dying(player) \rangle]$, which also corresponds to the dialectical tree rooted in $\langle \mathcal{A}_1, dying(player) \rangle$. Analogously, the only argumentation line starting with $\langle \mathcal{A}_2, \sim dying(player) \rangle$ (thus, the dialectical tree rooted in $\langle \mathcal{A}_2, \sim dying(player) \rangle$) is $[\langle \mathcal{A}_2, \sim dying(player) \rangle]$. As a result, argument $\langle \mathcal{A}_2, \sim dying(player) \rangle$ will be marked U on its dialectical tree and therefore, the answer to the query "dying(player)" is NO.*

As it is shown in the preceding example, the AI Director got a negative answer to the query "*dying(player)*", meaning that the player completed level 6 without risk of dying. Thus, the Director will follow the course of action leading to increase the difficulty of the next level, by reducing the available food and increasing the amount of enemies. The generation of level 7 is illustrated in Fig. 5, where the AI Director's knowledge base used to answer the query is visible. Observe that the amount of food & booze dropped on the floor in level 7 is 4, as a result of the AI Director's action (see Fig. 3) of subtracting 1 from the standard value of 5 (*i. e.*, the level number minus 2). Similarly, the AI Director increased the standard count of enemies (*i. e.*, the level number) by 1, resulting in 8 enemies to be faced in level 7.

Let us now consider a new scenario, similar to the one described in Example 1, where the only difference is that the player is about to complete level 12 instead of level 6. In this case, the AI Director will build its knowledge base and perform the corresponding queries before creating level 13. In particular, the knowledge base $\mathcal{P}_{12}$ will be such that it contains every rule and fact belonging to $\mathcal{P}_6$ plus the fact "*experienced(player)*", because the player is about to complete a level higher than 10 (*i. e.*, $\mathcal{P}_{12} = \mathcal{P}_6 \cup \{experienced(player)\}$).

When resolving the query "*dying(player)*" using $\mathcal{P}_{12}$, there exists a new counter-argument for $\langle \mathcal{A}_2, \sim dying(player)\rangle$: $\langle \mathcal{A}_3, \sim willRevive(player)\rangle$, with $\mathcal{A}_3 = \{\sim willRevive(player) \prec revivalAbility(player), experienced(player)\}$. In particular, $\langle \mathcal{A}_3, \sim willRevive(player)\rangle$ attacks the intermediate conclusion "*willRevive(player)*" of argument $\langle \mathcal{A}_2, \sim dying(player)\rangle$. Hence, the attacked sub-argument of $\langle \mathcal{A}_2, \sim dying(player)\rangle$ is $\langle \mathcal{A}_4, willRevive(player)\rangle$, with $\mathcal{A}_4 = \{willRevive(player) \prec revivalAbility(player)\}$. Then, since by generalized specificity $\langle \mathcal{A}_3, \sim willRevive(player)\rangle$ is preferred to the attacked sub-argument $\langle \mathcal{A}_4, willRevive(player)\rangle$, it holds that $\langle \mathcal{A}_3, \sim willRevive(player)\rangle$ is a defeater for $\langle \mathcal{A}_2, \sim dying(player)\rangle$. Moreover, since the only counter-argument for $\langle \mathcal{A}_3, \sim willRevive(player)\rangle$ is $\langle \mathcal{A}_4, willRevive(player)\rangle$, argument $\langle \mathcal{A}_3, \sim willRevive(player)\rangle$ is undefeated. As a result, the dialectical tree rooted in $\langle \mathcal{A}_1, dying(player)\rangle$ will correspond to the argumentation line $[\langle \mathcal{A}_1, dying(player)\rangle, \langle \mathcal{A}_2, \sim dying(player)\rangle, \langle \mathcal{A}_3, \sim willRevive(player)\rangle]$. Finally, argument $\langle \mathcal{A}_1, dying(player)\rangle$ will be marked U in its dialectical tree and, consequently, the answer to the query "*dying(player)*" will be *YES*.

As the preceding example shows, the smallest change in a scenario (in this case, the level number) may lead to different answers for the same query (thus, different actions of the AI Director), reflecting the defeasible nature of the information being represented within the AI Director's knowledge base.

5   Argumentative AI Director: Design and Implementation

In Unity, a DeLP Entity is represented by any *Prefab* or *GameObject* composed by a *custom script* that extends the *DeLP Entity script*. The DeLP Entity script is an abstract class that models the notion of DeLP Entity presented in Sect. 4. Thus, its attributes will be the sets of facts, strict rules, defeasible rules and queries of the entity. These attributes are initialized from the *Inspector*, allowing

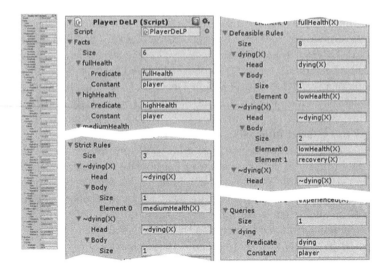

Fig. 6. Initialization of the Player Entity's attributes using the Inspector.

the user to abstract from their internal representation in the knowledge base. For instance, Fig. 6 illustrates how the *Player* Entity's attributes are initialized.

The DeLP Entity script will also be in charge of handling insertions and queries to the knowledge base. As explained in Sect. 4, a DeLP Entity may associate some conditions to its facts and queries, and also determines the actions to be performed depending on the queries' answers. Then, these conditions and actions will be specified within the custom script that implements the corresponding DeLP Entity script's abstract methods.

The knowledge base is part of another class implementing the *singleton* design pattern, allowing it to be globally accessible and restricting its instantiation to exactly one object. In the case of HermitArg, this is the same class containing a reference to every DeLP Entity, and is in charge of calling their methods before generating a new level. The classes involved within the design of HermitArg's AI Director are illustrated in Fig. 7.

In order to avoid involving the AI Director in unnecessary aspects, the methods used to handle and support the knowledge base are implemented inside the Java class *DeLPHandler*. This class provides us with the following methods:

- *addFact*: Adds a fact to the knowledge base.
- *addStrictRule*: Adds a strict rule to the knowledge base.
- *addDefeasibleRule*: Adds a defeasible rule to the knowledge base.
- *empty*: Empties the knowledge base.
- *query*: Makes a query to the knowledge base and returns a string with the corresponding answer: { *YES, NO, UNDECIDED, UNKNOWN* }.
- *toString*: For displaying purposes.

The above mentioned methods were implemented using the DeLP library provided by *Tweety* [15]. Briefly, Tweety is an open source project for scientific

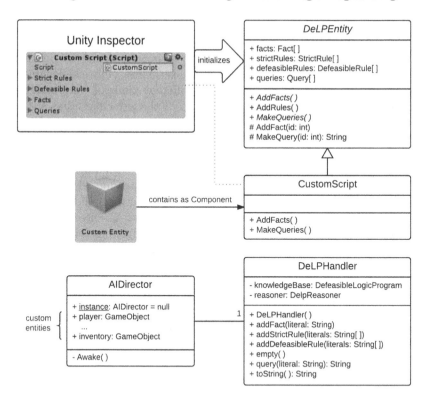

Fig. 7. Class diagram of HermitArg's AI Director.

experimentation on logical aspects of artificial intelligence and, particularly, knowledge representation. It provides a unified framework for implementing and testing knowledge representation formalisms, ranging from propositional logic to computational models of argumentation. Each formalism has a dedicated Java library that provides implementations for both syntactic and semantic constructs of the given formalism, as well as reasoning capabilities.

In particular, some of the classes provided by Tweety's DeLP library, which were used to implement the higher level methods of the *DeLPHandler* class, are: *DefeasibleLogicProgram*, which models a *DeLP program*; *DelpReasoner*, which implements the dialectical process carried out for the resolution of queries; *GeneralizedSpecificity*, which implements the homonym argument comparison criterion; *StrictRule*, which models a strict rule; *DefeasibleRule*, which models a defeasible rule; and *DelpFact*, which represents a literal modelling a fact. Finally, the resulting *.JAR* file encapsulating the implementation of the *DeLPHandler* is converted into a *.DLL* file, allowing for its use within the Unity project.

6   Related Work

In the last decade, there has been keen interest in addressing the problem of adaptivity in computer games in order to make games more challenging and appealing. In [10] the authors remark that different kinds of computer games may have different *purposes* of adaptivity (*i. e.*, the principles that steer the adaptation engine). For instance, adaptive entertainment games have typically only been considering one dimension to engage fun: challenge, meaning that the difficulty of performing game tasks must be in balance with the skills of the player avoiding undesirable "too easy" or "too hard" situations. In contrast, for educational or training games, the motivation for steering adaptability is to improve the effectiveness of the knowledge transfer between the game and its players. Since HermitArg is an entertainment game, its AI Director focuses on providing a fair and balanced level of challenge by adjusting the level of difficulty. Focusing on player-centered game adaptivity, our approach adjusts game elements depending on the individual performance of the player. Moreover, given that HermitArg is a basic roguelike game, the AI Director does not adjust level difficulty depending on the player's actions *per se*, but as a result of the overall behavior throughout the previous level. In contrast, for instance, in interactive narrative games each one of the player's actions and decisions have a direct impact on the direction or outcome of the storyline.

As mentioned in [13], a key challenge in interactive narrative systems is to maintain a coherent story progression while keeping users engaged. Hence, they implement an experience manager that drives the narrative forward by intervening in the fictional world, typically by directing computer-controlled characters in how to respond to the user's actions. Keeping the *coherence* goal in mind, the use of an Argumentative AI Director (experience manager) in this kind of games has a clear advantage. Argumentation systems are designed to deal with potentially contradictory information, and their underlying defeasible reasoning process provides the means to resolve those inconsistencies. Thus, the use of argumentation may come in handy for dealing with the *boundary problem* [11].

Another interesting approach to adaptive entertainment games is proposed in [7]. There, the authors present *Polymorph*, a 2-D platformer game that generates levels, as the user plays, driven by a dynamic difficulty adjustment to the player's skills and experience. A statistical model of difficulty and a model of the player's current skill level are used, through mass data collection and machine learning techniques, to select the appropriate level segments to generate for each user in order to avoid difficulty-related player frustration and boredom. It could be noted that, since the approach of [7] relies on human players' data collection for building the statistical model, adding a new game mechanic would make the old data obsolete. In contrast, by using an Argumentative AI Director, the incorporation of new elements in the game would only require to characterize new DeLP Entities and specify the facts, rules and queries associated with them. In addition, as mentioned by the authors in [7], their approach does not capture all aspects of level challenge because it focuses on the micro-level of component combinations (short segments of level and enemies) rather than level-wide patterns or the introduction of new mechanics. On the other hand, as mentioned before, our argumentative AI Director accounts for every aspect associated with the difficulty of the levels before generation of the next level.

A case study for a prototype of an adaptive *first-person shooter* gaming environment is presented in [8]. Player actions are recognized through a finite state machine approach, by which discrete actions reveal the player skill level. Adaptation mechanisms try to make the game harder for players identified as experts and easier for beginners in order to significantly increase player engagement, satisfaction, and ultimately enhance the game experience provided. Currently the recognition algorithm is composed from a variety of functionally independent mechanisms that are each executed in response to particular player initiated actions. For instance, the *"Kill Zone Counter"* mechanism enables the NPC populace to recognize when the player stands and snipes from a particular point using minimal movement to overcome NPCs, allowing the game to dispatch NPCs through an alternative route providing more of a challenge to the player. In [8] the authors say it would be interesting to develop a further level of function that observes the activation of each of these particular component mechanisms and attempts to autonomously weight each in relation to player progression throughout the level. This goal could be achieved by adding an Argumentative AI Director to the game, in which each DeLP Entity is in charge of executing its corresponding mechanisms and adding certain facts to the knowledge base depending on the new player status. Then, defeasible rules associated with each DeLP Entity could be used to argue whether the executed mechanism caused the desired effect and later adjust the corresponding parameters.

7   Conclusions and Future Work

In this paper we have presented a novel implementation of an AI Director for HermitArg, a roguelike video game developed in Unity 2D. Differently from other AI Directors in similar games, the decision making of our AI Director is supported by a defeasible argumentation process. For this purpose, we use the knowledge representation and reasoning formalism of DeLP.

Before generating a new level, the AI Director builds a knowledge base in the form of a DeLP program, representing information about the status of different entities in the game (player, inventory, etc.) at the end of the previous level. Then, the Director performs some queries on the knowledge base, which are resolved using DeLP's dialectical process, and adopts a different course of action depending on the obtained answers. Specifically, these answers may lead the Director to increase or decrease the difficulty of the following level, in order to provide a fair and balanced level of challenge while keeping the player interested.

The use of defeasible logic allows the formalization of *smart items* in the game scenario. A smart item, or technically *DeLP Entity*, is a level object with an associated set of logic rules, stating defeasible conditions about the adaptability of such an object. The abstraction of smart items leads to a modular design about the overall knowledge regarding a single level. This modular design adopted by HermitArg's AI Director has two main advantages. On the one hand, since the sets of facts and rules associated with each DeLP Entity is a DeLP program on its own, the resulting AI Director's knowledge base is a combination of such programs. This allows the AI Director to abstract from the presence (or absence) of the different entities at each moment in the game, by simply asking the available entities to incorporate their DeLP programs into the knowledge base. At any case, this would only affect the resolution of queries, as explained below. For instance, let us suppose that the Player is not allowed to access the Inventory at a given point in the game. Then, the DeLP program corresponding to the *Player* entity would be added to the Director's knowledge base, whereas the one associated with the *Inventory* would not. The resulting knowledge base would still be a valid DeLP program. However, since some rules of the Player entity make use of information added by the Inventory entity, it would not be possible to build the arguments that contain them. As a result, the answer to the queries of the Player entity may change. On the other hand, the modular design and the high level of sophistication of DeLP's defeasible argumentation formalism allow for the specification of complex rules within each entity. Then, since the resolution of queries is carried out by the DeLP reasoner, the AI Director can abstract from the level of complexity of the dialectical process, which may involve the consideration of a wide range of arguments and conflicts.

Future work has several directions. We are interested in applying the notion of argumentative AI Director to first-person shooter games such as *Doom 3* or *Quake III Arena*. Because of the fast-paced environment of these games, the incorporation of an argumentation formalism like DeLP into their AI Director poses a great challenge. In that way, for instance, AI Director will have to constantly update and query the knowledge base in order to decide the actions required to dynamically adapt and change scenarios depending on the player's performance. A small bot testing the use argumentation in very dynamic scenarios was developed using *TORCS: The Open Car Racing Simulator* and the libraries for *Demolition Derby* 2012. We are also working on the use of DeLP in the creation of bots that use argumentation to decide about their actions, using a knowledge base obtained by crowdsourcing and the analysis of contradictory and incomplete game logs.

References

1. Besnard, P., Hunter, A.: A logic-based theory of deductive arguments. Artif. Intell. **128**(1–2), 203–235 (2001)
2. Booth, M.: The AI systems of Left 4 Dead. In: Keynote, 5th AIIDE (2009)
3. DeLP Web page. http://lidia.cs.uns.edu.ar/delp
4. Dung, P.M.: On the acceptability of arguments and its fundamental role in non-monotonic reasoning, logic programming and n-person games. Artif. Intell. **77**(2), 321–358 (1995)
5. García, A.J., Simari, G.R.: Defeasible logic programming: an argumentative approach. Theor. Pract. Logic Program. **4**(1–2), 95–138 (2004)
6. Harrison, B.E., Roberts, D.L.: Analytics-driven dynamic game adaption for player retention in a 2-dimensional adventure game. In: 10th AIIDE, pp. 23–29 (2014)

7. Jennings-Teats, M., Smith, G., Wardrip-Fruin, N.: Polymorph: dynamic difficulty adjustment through level generation. In: Proceedings of 2010 Workshop on Procedural Content Generation in Games, pp. 1–14 (2010)
8. Kazmi, S., Palmer, I.J.: Action recognition for support of adaptive gameplay: a case study of a first person shooter. Int. J. Comp. Games Tech. **2010**, 1–14 (2010)
9. Lifschitz, V.: Foundations of logic programs. In: Brewka, G. (ed.) Principles of Knowledge Representation, pp. 69–128. CSLI Pub. (1996)
10. Lopes, R., Bidarra, R.: Adaptivity challenges in games and simulations: a survey. IEEE Trans. Comput. Intell. AI Games **3**(2), 85–99 (2011)
11. Magerko, B.: Evaluating preemptive story direction in the interactive drama architecture. J. Game Dev. **2**(3), 25–52 (2005)
12. Rahwan, I., Simari, G.R.: Argumentation in Artificial Intelligence. Springer, New York (2009)
13. Riedl, M.O., Bulitko, V.: Interactive narrative: an intelligent systems approach. AI Mag. **34**(1), 67–77 (2013)
14. Talbot, C.: Creating an artificially intelligent director (aid) for theatre and virtual environments. In: 12th AAMAS, pp. 1457–1458 (2013)
15. Thimm, M.: Tweety: A comprehensive collection of java libraries for logical aspects of artificial intelligence and knowledge representation. In: 14th International Conference on Principles of Knowledge Representation and Reasoning, pp. 528–537 (2014)

General Intelligence in Game-Playing Agents 2015

On the Cross-Domain Reusability of Neural Modules for General Video Game Playing

Alex Braylan, Mark Hollenbeck, Elliot Meyerson[✉], and Risto Miikkulainen

Department of Computer Science, The University of Texas at Austin, Austin, USA
{braylan,mhollen,ekm,risto}@cs.utexas.edu

Abstract. We consider a general approach to knowledge transfer in which an agent learning with a neural network adapts how it reuses existing networks as it learns in a new domain. Networks trained for a new domain are able to improve performance by selectively routing activation through previously learned neural structure, regardless of how or for what it was learned. We consider a neuroevolution implementation of the approach with application to reinforcement learning domains. This approach is more general than previous approaches to transfer for reinforcement learning. It is domain-agnostic and requires no prior assumptions about the nature of task relatedness or mappings. We analyze the method's performance and applicability in high-dimensional Atari 2600 general video game playing.

1 Introduction

The ability to generally apply any and all available previously learned knowledge to new tasks is a hallmark of general intelligence. *Transfer learning* is the process of reusing knowledge from previously learned *source* tasks to bootstrap learning of *target* tasks. For reinforcement learning (RL) agents, transfer is particularly important, as previous experience can help to efficiently explore new environments. Knowledge acquired during previous tasks also contains information about the agent's task-independent decision making and learning dynamics, and thus can be useful even if the tasks seem completely unrelated.

Existing approaches to transfer learning for reinforcement learning have successfully demonstrated transfer of varying kinds of knowledge [24], but they tend to make two fundamental assumptions that restrict their generality: (1) some sort of a priori human-defined understanding of how tasks are related, (2) separability of knowledge extraction and target learning. The first assumption limits the applicability of the approach by restricting its use only to cases where the agent has been provided with this additional relational knowledge, or, if it can be learned, cases where task mappings are useful. The second assumption implies further expectations about what knowledge will be useful and how it should be incorporated *before* learning on the target task begins, preventing the agent from adapting the way it uses source knowledge as it gains information about the target domain.

© Springer International Publishing Switzerland 2016
T. Cazenave et al. (Eds.): CGW 2015/GIGA 2015, CCIS 614, pp. 115–129, 2016.
DOI: 10.1007/978-3-319-39402-2_9

We consider General ReUse Of Static Modules (GRUSM), a general neural network approach to transfer learning that avoids both of these assumptions, by augmenting the learning process to allow learning networks to selectively route through existing neural modules (source networks) as they simultaneously develop new structure for the target task. Unlike previous work, which has dealt with mapping task variables between source and target, GRUSM is task-independent, in that no knowledge about the structure of the source task or even knowledge about where the network came from is required for it to be reused. Instead of using mappings between task-spaces to facilitate transfer, it searches directly for mappings in the solution space, that is, connections between existing source networks and the target network. GRUSM is motivated by studies that have shown in both naturally occurring complex networks [14] and artificial neural networks [21] that certain network structures repeat and can be useful across domains, without any context for how exactly this structure should be used. We are further motivated by the idea that neural resources in the human brain are reused for countless purposes in varying complex ways [1].

In this paper, we consider an implementation of GRUSM based on the Enforced Subpopulations (ESP) neuroevolution framework [6]. We validate our approach first in a simple boolean logic domain, then scale up to the Atari 2600 general game playing domain. In both domains, we find that GRUSM-ESP improves learning overall, and tends to be most useful when source and target networks are more complex. In the Atari domain, we show that the effectiveness of transfer coincides with an intuitive high-level understanding of game dynamics. This demonstrates that even without traditional transfer learning assumptions, successful knowledge transfer via general reuse of existing neural modules is possible and useful for RL. In principle, our approach and implementation naturally scale to transfer from an arbitrary number of source tasks, which points towards a future class of GRUSM agents that accumulate and reuse knowledge throughout their lifetimes across a variety of diverse domains.

The remainder of this paper is organized as follows: Sect. 2 provides background on transfer learning and related work, Sect. 3 describes our approach in detail, Sect. 4 analyzes results from experiments we have run with this approach, and Sect. 5 discusses the implications of these results and motivations for future work.

2   Background

Transfer learning encompasses machine learning techniques that involve reusing knowledge across different domains and tasks. In this section we review existing transfer learning methodologies and discuss their advantages and shortcomings to motivate our approach. We take the following two definitions from [16]. A *domain* is an environment in which learning takes place, characterized by the input and output space. A *task* is a particular function from input to output to be learned. In sequential-decision domains, a task is characterized by the values of sensory-action sequences corresponding to the pursuit of a given goal.

A taxonomy of types of knowledge that may be transferred are also enumerated in [16]. As our approach reuses the structure of existing neural networks, it falls under 'feature representation transfer'.

2.1 Transfer Learning for RL

Reinforcement learning (RL) domains are often formulated as Markov decision processes in which the state space comprises all possible observations, and the probability of an observation depends on the previous observation and an action taken by a learning agent. However, many real world RL domains are non-Markovian, including many Atari 2600 games.

Five dimensions for characterizing the generality and autonomy of algorithms for transfer learning in RL are given in [24]: (1) restrictions on how source and target task can differ; (2) whether or not *mappings* between source and target state and action variables are available to assist transfer; (3) the form of the knowledge transferred; (4) restrictions on what classes of learning algorithms can be used in the source and/or target tasks; (5) whether or not the algorithm autonomously selects which sources to reuse.

Some of the most general existing approaches to transfer for RL automatically learn task mappings, so they need not be provided beforehand, e.g., [22,23,25]. These approaches are general enough to apply to any reinforcement learning task, but as the state and action spaces become large they become intractable due to combinatorial blowup in the number of possible mappings. These approaches also rely on the assumption that knowledge for transfer can be extracted based on mappings between state and action variables, which may miss useful internal structure these mappings cannot capture.

2.2 General Neural Structure Transfer

There are existing algorithms similar to our approach in that they enable general reuse of existing neural structure. They can apply to a wide range of domains and tasks in that they automatically select source knowledge and avoid inter-task mappings. Knowledge-Based Cascade Correlation [20] uses a technique based on cascade correlation to build increasingly complex networks by inserting source networks chosen by how much they reduce error. Knowledge Based Cascade Correlation is restricted in that it is only designed for supervised learning, as the source selection depends heavily on an immediate error calculation. Also, connectivity between source and target networks is limited to the input and output layer of the source. Subgraph Mining with Structured Representations [21] creates sparse networks out of primitives, or commonly used sub-networks, mined from a library of source networks. The subgraph mining approach depends on a computationally expensive graph mining algorithm, and it tends to favor exploitation over innovation and small primitives rather than larger networks as sources.

Our approach is more general in that it can be applied to unsupervised and reinforcement learning tasks, and makes fewer a priori assumptions about what kind of sources and mappings should work best. Although we only consider an evolutionary approach in this paper, GRUSM should be extensible to any neural network-based learning algorithm.

3   Approach

This section introduces the general idea behind GRUSM, then provides an overview of the ESP neuroevolution framework, before describing our particular implementation: GRUSM-ESP.

3.1   General ReUse of Static Modules (GRUSM)

The underlying idea is that an agent learning a neural network for a target task can selectively reuse knowledge from existing neural modules (source networks) while simultaneously developing new structure unique to a target task. This attempts to balance reuse and innovation in an integrated architecture. Both source networks and new hidden nodes are termed *recruits*. Recruits are added to the target network during the learning process. Recruits are adaptively incorporated into the target network as it learns connection parameters from the target to the recruit and from the recruit to the target. All internal structure of source networks is *frozen* to allow learning of connection parameters to remain consistent across recruits. This forces the target network to transfer learned knowledge, rather than simply overwrite it. Connections to and from source networks can, in the most general case, connect to any nodes in the source and target, minimizing assumptions about what knowledge will be useful.

A GRUSM network or *reuse network* is a 3-tuple $\mathcal{G} = (M, S, T)$ where M is a traditional neural network (feedforward or recurrent) containing the new nodes and connections unique to the target task, with input and output nodes corresponding to inputs and outputs defined by the target domain; S is a (possibly empty) set of pointers to recruited source networks $\mathcal{S}_1, ..., \mathcal{S}_k$; and T is a set of weighted *transfer connections* between nodes in M and nodes in source networks, that is, for any connection $((u, v), w) \in T$, $(u \in M \land v \in \mathcal{S}_i) \lor (u \in \mathcal{S}_i \land v \in M)$ for some $0 \leq i \leq k$. This construction strictly extends traditional neural networks so that each $\mathcal{S}_i$ can be a traditional neural network or a reuse network of its own. When $\mathcal{G}$ is evaluated, we evaluate only the network induced by directed paths from inputs of M to outputs of M, including those which pass through some $\mathcal{S}_i$ via connections in T. Before each evaluation of $\mathcal{G}$, all recruited source network inputs are set to 0, since at any given time the agent is focused only on performing the current target task. The parameters to be learned are the usual parameters of M, along with the contents of S and T. The internal parameters of each $\mathcal{S}_i$ are frozen in that they cannot be rewritten through $\mathcal{G}$.

The motivation for this architecture is that if the solution to a source task contains *any* information relevant to solving a target task, then the neural network constructed for the source task will contain *some* structure (subnetwork or module) that will be useful for a target network. This has been shown to be true in naturally occurring complex networks [1], as well as cross-domain artificial neural networks [21]. Unlike in the subgraph mining approach to neural structure transfer [21], this general formalism makes no assumptions as to what subnetworks actually will be useful. One perspective that can be taken with this approach is that a lifelong learning agent maintains a system of interconnected neural modules that it can potentially reuse at any time for a new task. Even if existing modules are unlabeled, they may still be useful, simply due to the fact that they contain knowledge of how the agent can successfully learn. Furthermore, recent advances in reservoir computing [12] have demonstrated the power of using large amounts of frozen neural structure to facilitate learning of complex and chaotic tasks.

The above formalism is general enough to allow for an arbitrary number of source networks and arbitrary connectivity between source and target. In this paper, to validate the approach and simplify analysis, we use at most one source network and only allow connections from target input to source hidden layer and source output layer to target output. This is sufficient to show that the implementation can successfully reuse hidden source features, and analyze the cases in which transfer is most useful. Future refinements are discussed in Sect. 5. The current implementation, described below, is a neuroevolution approach based on ESP.

3.2 Enforced Subpopulations (ESP)

Enforced Sub-Populations (ESP) [6] is a neuroevolution technique in which different elements of a neural network are evolved in separate *subpopulations* rather than evolving the whole network in a single population. ESP has been shown to perform well in a variety of reinforcement learning tasks, e.g., [6–8,13,19]. In standard ESP, each hidden neuron is evolved in its own subpopulation. Recombination occurs only between members of the same subpopulation, and mutants in a subpopulation derive only from members of that subpopulation. The genome of each individual in a subpopulation is a vector of weights corresponding to the weights of connections from and to that neuron, including node bias. In each generation, networks to be evaluated are randomly constructed by inserting one neuron from each subpopulation. Each individual that participated in the network receives the fitness achieved by that network. When fitness converges, i.e., does not improve over several consecutive generations, ESP makes use of *burst phases*. In initial burst phases each subpopulation is repopulated by mutations of the single best neuron ever occurring in that subpopulation, so that it reverts to searching a δ-neighborhood around the best solution found so far. If a second consecutive burst phase is reached, i.e., no improvements were made since the previous burst phase, a new neuron with a new subpopulation may be added [5].

3.3 GRUSM-ESP

We extend the idea of enforced sub-populations to transfer learning via GRUSM networks. For each reused source network $\mathcal{S}_i$ the transfer connections in T

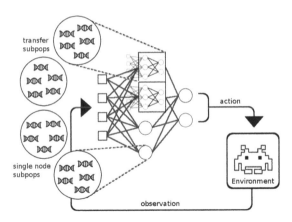

Fig. 1. The GRUSM-ESP architecture. Reused subnetworks of sources are boxed. Edges between input and source and between source and output denote full connectivity between these layers. The genome in each subpopulation encodes weight information for the connections from and to the corresponding recruit.

between $\mathcal{S}_i$ and M evolve in a distinct subpopulation. At the same time new hidden nodes can be added to M and evolve within their own subpopulations in the manner of standard ESP. In this way, the integrated evolutionary process simultaneously searches the space for how to reuse each potential source network and how to innovate with each new node. Specifically, the GRUSM-ESP architecture (Fig. 1) is composed of the following elements:

– A pool of potential source networks. In the experiments in this paper, each target network reuses at most one source at a time.
– *Transfer genomes* defining a list of transfer connections between the source and target networks. Each potential source network in the pool has its own subpopulation for evolving transfer genomes between it and the target network. Each connection in T is contained in some transfer genome. In our experiments, the transfer connections included are those such that the target's inputs are fully connected to the source's hidden layer, and the source's outputs are fully connected into the target's outputs. Therefore, the transfer genome only encodes the weights of these cross-network connections.
– A burst mechanism that determines when innovation is necessary based on a recent history of performance improvement. New hidden recruits (source networks or new single nodes) added during the burst phase evolve within their own subpopulations in the manner of classic ESP.

All hidden and output neurons use a hyperbolic tangent activation function. Networks include a single hidden layer, and can include self-loops on hidden nodes; they are otherwise feedforward. The particulars of the genetic algorithm in our implementation used to evolve each subpopulation mirror those described in [5]. This algorithm has been shown to work well within the ESP framework, though any evolutionary algorithm could potentially be substituted in its place

Table 1. Median number of generations for task completion for all N-bit parity source-target setups.

	Source	None	2-bit parity	3-bit parity	4-bit parity
Target	3-bit parity	309.5	202	167.5	**158**
	4-bit parity	339	**192.5**	308	311
	5-bit parity	626	780	720.5	**542**

4   Experiments

We evaluate GRUSM-ESP on two domains: a simple n-bit parity domain mirroring that used to evaluate knowledge transfer in [21], and the more complex Atari 2600 video game playing domain. We first train *scratch* networks that do not reuse existing networks, that is, S is the empty set. We then reuse each scratch network in training GRUSM networks for different tasks. We compare performance between scratch and transfer, and between source-target setups. Results demonstrate the ability of GRUSM-ESP to selectively reuse source structure.

4.1   N-Bit Parity

GRUSM-ESP was initially evaluated under the boolean logic domain using N-bit parity. The N-bit parity problem has a long-standing history serving as a benchmark for basic neural network performance. The N-bit parity function is the mapping defined on N-bit binary vectors that returns 1 if the sum of the N binary values in the vector is odd, and 0 otherwise. This function is deceptively difficult for neural networks to learn since a change in any single input bit will alter the output. Although N-bit parity is not fully *cross-domain* in the stronger sense for which our approach applies, the input feature space does differ as N differs, and it is useful for validation of the approach and connection with previous work.

Performance is measured in number of generations to find a network that solves N-bit parity within $\epsilon = 0.1$ mean squared error. In this experiment, networks were trained from scratch with ESP for $N = [2, 3, 4]$. Then, each of these networks was used as a source network for each N-bit parity target domain with $N = [3, 4, 5]$. ESP, without transfer, was used as a control condition for each target task. A total of 10 trials were completed for each condition.

In this experiment, transfer learning was able to outperform learning from scratch for all three target tasks when using some source task (Table 1). For 3-bit and 4-bit parity, transfer learning always outperformed learning from scratch for all three possible sources. For the more complex 5-bit parity target task, transfer from the 4-bit network outperformed learning from scratch, while transfer from the simpler tasks did not. This may be due to the significantly greater complexity required for 5-bit parity over 2- or 3-bit parity. The limited frozen structure may become a burden to innovation after the initial stages of evolution. The more complex 4-bit parity networks have more structure to select from, and thus may assist in innovation over a longer time frame.

4.2   Atari 2600 Game Playing

Our next experiment evaluated in the Atari 2600 game platform using the Arcade Learning Environment (ALE) simulator [2]. This domain is particularly popular for evaluating RL techniques, as it exhibits sufficient complexity to challenge modern approaches, contains non-Markovian properties, and entertained a generation of human video game players. We used GRUSM-ESP to train agents to play eight games (Asterix, Bowling, Boxing, Breakout, Freeway, Pong, Space Invaders, and Seaquest) both from scratch and using transferred knowledge from existing game-playing source networks. Neuroevolution techniques are quite competitive in the Atari 2600 domain [9], and ESP in particular has yielded state-of-the-art performance for several games [3].

Each source network was trained from scratch on a game using standard ESP (GRUSM-ESP with an empty reuse set). Each source network was then used by a target network for an evolutionary run for each other game. Each run lasted 200 generations with 100 evaluations per generation. Each individual i achieves some score $i(g)$ in its game g. Let $min(g)$ be the min over all max scores achieved in a single generation by any run of g. Let the *fitness* of i be $i(g) - min(g)$. This ensures that fitness is always positive (in both boxing and pong, raw scores can be negative). The fitness of an evolutionary run at a given generation is the highest fitness achieved by an individual by that generation.

We ran a total of 176 trials split across all possible setups: training using each other game as a source, and training from scratch. We use the ϵ-repeat action approach as suggested in [10] to make the environment stochastic in order to disable the algorithm from finding loopholes in the deterministic nature of the simulator. We use the recommended $\epsilon = 0.25^1$. Parameters were selected based on their success with standard ESP.

To interface with ALE, the output layer of each network consists of a 3×3 substrate representing the 9 directional movements of the Atari joystick in addition to a single node representing the Fire button. The input layer consists of a series of object representations manually generated as previously described in [9], where the location of each object on the screen is represented in an 8×10 input substrate corresponding to the object's class. The number of object classes for the games used in our experiments varies between one and four. Although object representations are used in these experiments, pixel-level vision could also be learned from scratch below the neuroevolution process, e.g., via convolutional networks, as in [11].

Domain Characterization. Each game can be characterized by generic binary features that determine the requirements for successful game play, in order to

---

[1] https://github.com/mgbellemare/Arcade-Learning-Environment/tree/dev.

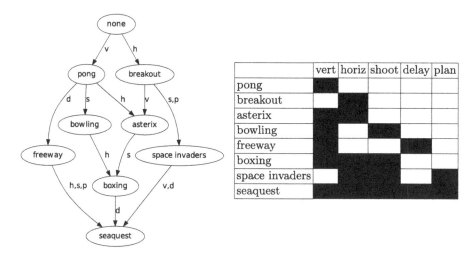

Fig. 2. (left) Lattice of games ordered by features, and (right) vector of features for each game (indicated in black). Every path from *none* to *g* contains along its edges each complexity feature of *g* exactly once. Features: v = vertical movement, h = horizontal movement, s = shooting, d = delayed reward, p = long-term planning.

	vert	horiz	shoot	delay	plan
pong	■				
breakout		■			
asterix	■	■			
bowling	■		■		
freeway	■			■	
boxing	■	■	■		
space invaders		■	■		■
seaquest	■	■	■	■	■

place the games within a unified framework. We use binary features based on the existence of the following: (1) horizontal movement (joystick left/right), (2) vertical movement (joystick up/down), (3) shooting (fire button); (4) delayed rewards; and (5) the requirement of long-term planning. Intuitively, more complex games will possess more of these qualities. A partial ordering of games by complexity defined by these features is shown in Fig. 2. The assignment of features (1), (2) and (3) is completely defined based on game interface [2]. Freeway and Seaquest are said to have *delayed rewards* because a high score can only be achieved by long sequences of rewardless behavior. Only Space Invaders and Seaquest were deemed to require long-term planning [15], since the long-range dynamics of these games penalize reflexive strategies, and as such, agents in these games can perform well with a low decision-making frequency [3]. Aside from their intuitiveness, these features are validated below based on their ability to characterize games by complexity and predict transferability. For a simple metric of complexity, let $cmplx(g)$ be the number of the above features game g exhibits.

Atari 2600 Results. There are many possible approaches to evaluating success of transfer [24]. For comparing performance *across* games, we focus on *time to threshold*. To minimize threshold bias, for each game we chose the threshold to be the min of the max fitness achieved across all trials. Given this threshold, the average time to threshold in terms of generations may be vastly different, depending on the average learning curve of each game. These learning curves are quite irregular, as illustrated in Fig. 3. For each game we measure time in terms

of percent of average time to threshold, and the *success rate* is the proportion
of trials that have achieved the threshold by that time.

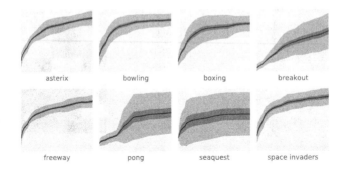

Fig. 3. Distributions of fitness for each game by generation over all trials. Mean (black),
standard error (dark gray) and standard deviation (light gray) are shown at each
generation.

Figure 4 plots success over time for different groups of trials. The top plot
compares the success rate of all transfer trials to scratch trials. It shows what
we would expect from transfer overall: networks that reuse frozen structure from
previous games are able to take advantage of that structure to bootstrap learning. This works initially, but eventually scratch catches up, as it becomes more
difficult to innovate with a single frozen structure. When trials are grouped by
target (lower left pane), we can see that some games are better targets for transfer than others. As demonstrated in Fig. 4, more complex games (with respect to
our game features) are generally better targets than less complex. It is less clear
what we can draw from grouping trials by source (lower right pane). There is a
tighter spread than with targets, though there may still be a tendency towards
more complex games being better sources. This may be counter-intuitive, as
we might expect simpler games to be easier to reuse. However, more complex
games have networks with more complex structure from which a target network
can, through the evolutionary process, select some useful subnetwork that fits
its needs. Similarly, a complex domain will be more likely to be a good target,
since it requires a wider variety of structure to be successful, so sources have a
higher chance of satisfying *some* of that requirement.

For comparing performance *within* a target game, we need not resort to
threshold normalization, and can instead focus on raw max fitness. For reference, average and best fitness for both transfer and scratch are given in Table 2.
Note that previously published approaches to Atari game-playing use fully deterministic environments, making direct score comparisons difficult (see [3] for a
comparison of ESP to other approaches in deterministic environments).

The *max fitness transfer effectiveness* (MFTE) of a source-target setup is
the log ratio between its average max fitness and the average max fitness of
that game from scratch. The digraphs in Fig. 5 each contain the directed edge

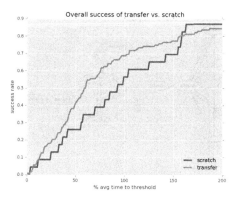

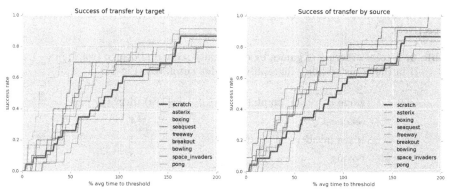

Fig. 4. Success rate (proportion of trials that have reached the target threshold) by percent of average number of generations to threshold of target game with trials (allowing comparisons across games with different average times to threshold) grouped by (1) scratch vs. transfer, (2) target game, (3) source game.

Table 2. For both transfer (t) and scratch (s) runs, average fitness and best fitness of GRUSM-ESP.

game	$min(g)$	$best_t$	$best_s$	avg_t	avg_s
seaquest	160	1510	300	475.0	262.0
space invaders	310	1520	1320	1076.0	1160.0
boxing	−12	111	107	98.6	104.1
bowling	30	237	231	219.9	201.9
asterix	650	3030	2150	1989.0	2016.7
freeway	21	13	11	10.7	10.7
pong	−21	42	42	21.8	20.3
breakout	0	51	37	25.4	31.3

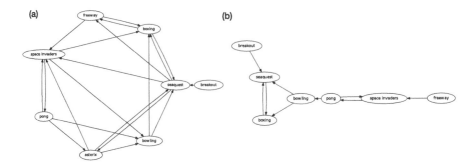

Fig. 5. Transferability graphs illustrating the most successful source-target pairs. Each graph includes a directed edge from g_1 to g_2 $\Longleftrightarrow$ the MFTE (max fitness transfer effectiveness; defined above) for g_2 reusing g_1 is greater than (a) 0.5, and (b) 1.0, respectively.

Table 3. A total ordering of games by complexity score and degree (total, in ($-$), and out ($+$)) in the transferability digraph with edge cutoff 0.5 (Fig. 5(a)).

game g	cmplx(g)	deg(g)	deg$^-$(g)	deg$^+$(g)
seaquest	5	8	4	4
space invaders	3	7	4	3
boxing	3	6	4	2
bowling	2	5	3	2
asterix	2	5	2	3
freeway	2	4	2	2
pong	1	4	1	3
breakout	1	1	0	1

from g_1 to g_2 only when MFTE is above a specified threshold. These graphs indicate that the more complex games serve a more useful role in transfer than less complex. Consider the total ordering of games by $cmplx(g)$ given in Table 3. This ordering corresponds exactly to that induced by the degree sequence (by both total degree and in-degree) [4] of the graph with edge cutoff 0.5. However, for out-degree, the correlation with respect to the ordering is less clear. This reflects Fig. 4, in which there is more spread in success when grouped by target (in-degree) vs. source (out-degree).

We see that we can predict MFTE by the feature characterizations we provided. The feature characterizations allow us to consider all trials in the same feature space. A linear regression model trained on a random half of the setups yielded weight coefficients for the source and target features that successfully predicted the MFTE of setups in the test set (Fig. 6). The slope was found to be statistically significant with a p-value < 0.01. The most significant features were vertical movement and long-term planning in the source domain, with respective coefficients of 0.73 and 0.89. The ability to use the game features to predict

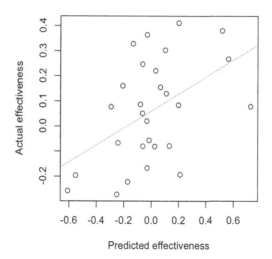

Fig. 6. Feature-based linear prediction versus actual MFTE (max fitness transfer effectiveness) on out-of-sample setups.

MFTE can be used to inform source selection. It is also encouraging that the effectiveness of transfer with GRUSM-ESP correlates with a high-level intuition of inter-game dynamics.

5 Discussion

Our results show that GRUSM-ESP, an evolutionary algorithm for general transfer of neural network structure, can improve learning in both boolean logic and Atari game playing by reusing previously developed knowledge. However, we find that the improvement in learning performance in the target domain depends heavily on the source network. Some source-target pairs do not consistently outperform training from scratch, indicating negative transfer from that source. This highlights the importance of source selection in transfer learning.

Specifically with the Atari game playing domain, we observe an issue of source knowledge *quality*. Some of the source networks that were trained from scratch do relatively well on games whereas others do not. One problem is that the measure of knowledge in source networks is ill-defined. As alluded to in [25], there could be an optimal point in a source's training at which to transfer knowledge to a target, after which the source network has encoded knowledge too specific to its own task, which does not generalize as well to other tasks, and makes useful knowledge difficult to extract. Future analysis will investigate topological regularities of source networks and transfer connections, to further address what and how knowledge is successfully reused.

Another future area of work will involve increasing the flexibility in the combined architecture by (1) relaxing the requirement for all transfer connections to be input-to-hidden and output-to-output, and (2) allowing deeper architectures for the source and target networks. This will promote reuse of subnetworks of varying depth and flexible positioning of modules. However, as networks become large and plentiful, maintaining full connectivity between layers will become intractable, and enforcing sparsity will be necessary.

Having shown that our algorithm works with certain target-source pairs, a next step will involve pooling multiple candidate sources and testing GRUSM-ESP's ability to exploit the most useful ones. GRUSM-ESP extends naturally to learning transfer connections for multiple sources simultaneously. By starting with limited connectivity and adding connections to sources that show promise (while removing connections from ones that are not helping), adaptive multi-source selection may be integrated into the evolutionary process. Methods for adapting this connectivity online have yet to be developed.

Although our initial experiments only scratched the surface, they are encouraging in that they show general transfer of neural structure is possible and useful. They have also helped us characterize the conditions under which transfer may be useful. It will be interesting to investigate whether the same principles extend to other general video game playing domains, such as [17,18]. This should help us better understand how subsymbolic knowledge can be recycled indefinitely across diverse domains.

6   Conclusion

We consider a framework for general transfer learning using neural networks. This approach minimizes a priori assumptions of task relatedness and enables a flexible approach to adaptive learning across many domains. In both the Atari 2600 and N-bit parity domains, we show that a specific implementation, GRUSM-ESP is able to successfully boost learning by reusing neural structure across disparate tasks. The success of transfer is shown to correlate with intuitive notions of task dynamics and complexity. Our results indicate that general neural reuse – a staple of biological systems – can effectively assist agents in increasingly complex environments.

Acknowledgments. This research was supported in part by NSF grant DBI-0939454, NIH grant R01-GM105042, and an NPSC fellowship sponsored by NSA.

References

1. Anderson, M.L.: Neural reuse: a fundamental organizational principle of the brain. Behav. Brain Sci. **33**, 245–266 (2010)
2. Bellemare, M.G., Naddaf, Y., Veness, J., Bowling, M.: The arcade learning environment: an evaluation platform for general agents. J. Artif. Intell. Res. **47**, 253–279 (2013)
3. Braylan, A., Hollenbeck, M., Meyerson, E., Miikkulainen, R.: Frame skip is a powerful parameter for learning to play atari. In: AAAI Workshop on Learning for General Competency in Video Games (2015)
4. Diestel, R.: Graph Theory, p. 278. Springer, Heidelberg (2005)
5. Gomez, F.J.: Robust non-linear control through neuroevolution. Technical report, UT Austin (2003)

6. Gomez, F.J., Miikkulainen, R.: Incremental evolution of complex general behavior. Adapt. Behav. **5.3**(5), 317–342 (1997)
7. Gomez, F.J., Miikkulainen, R.: Active guidance for a finless rocket using neuroevolution. In: Proceedings of GECCO 2003, pp. 2084–2095 (2003)
8. Gomez, F.J., Schmidhuber, J.: Co-evolving recurrent neurons learn deep memory pomdps. In: Proceedings of GECCO 2005, pp. 491–498 (2005)
9. Hausknecht, M., Lehman, J., Miikkulainen, R., Stone, P.: A neuroevolution approach to general atari game playing. In: Computational Intelligence and AI in Games (2013)
10. Hausknecht, M., Stone, P.: The impact of determinism on learning Atari 2600 games. In: AAAI Workshop on Learning for General Competency in Video Games (2015)
11. Koutník, J., Schmidhuber, J., Gomez, F.: Online evolution of deep convolutional network for vision-based reinforcement learning. In: del Pobil, A.P., Chinellato, E., Martinez-Martin, E., Hallam, J., Cervera, E., Morales, A. (eds.) SAB 2014. LNCS, vol. 8575, pp. 260–269. Springer, Heidelberg (2014)
12. Lukoševičius, M., Jaeger, H.: Reservoir computing approaches to recurrent neural network training. Comput. Sci. Rev. **3**(3), 127–149 (2009)
13. Miikkulainen, R., Feasley, E., Johnson, L., Karpov, I., Rajagopalan, P., Rawal, A., Tansey, W.: Multiagent learning through neuroevolution. In: Alippi, C., Bouchon-Meunier, B., Greenwood, G.W., Abbass, H.A., Liu, J. (eds.) WCCI 2012. LNCS, vol. 7311, pp. 24–46. Springer, Heidelberg (2012)
14. Milo, R., Shen-Orr, S., Itzkovitz, S., Kashtan, N., Chklovskii, D., Alon, U.: Network motifs: simple building blocks of complex networks. Science **298**(5594), 824–827 (2002)
15. Mnih, V., Kavukcuoglu, K., Silver, D., Graves, A., Antonoglou, I., Wierstra, D., Riedmiller, M.: Playing atari with deep reinforcement learning (2013). arXiv:1312.5602
16. Pan, S.J., Yang, Q.: A survey on transfer learning. Knowl. Data Eng. **22**(10), 1345–1359 (2010)
17. Perez, D., Samothrakis, S., Togelius, J., Schaul, T., Lucas, S., Couetoux, A., Lee, J., Lim, C., Thompson, T.: The 2014 general video game playing competition. IEEE Trans. Comput. Intell. AI Games (2015)
18. Schaul, T.: A video game description language for model-based or interactive learning. In: Proceedings of IEEE Conference on Computational Intelligence in Games (CIG 2013), pp. 193–200 (2013)
19. Schmidhuber, J., Wierstra, D., Gagliolo, M., Gomez, F.J.: Training recurrent networks by Evolino. Neural Comput. **19**(3), 757–779 (2007)
20. Shultz, T.R., Rivest, F.: Knowledge-based cascade-correlation: using knowledge to speed learning. Connection Sci. **13**(1), 43–72 (2001)
21. Swarup, S., Ray, S.R.: Cross-domain knowledge transfer using structured representations. In: Proceedings of AAAI, pp. 506–511 (2006)
22. Talvitie, E., Singh, S.: An experts algorithm for transfer learning. In: Proceedings of IJCAI 2007, pp. 1065–1070 (2007)
23. Taylor, M.E., Kuhlmann, G., Stone, P.: Autonomous transfer for reinforcement learning. In: Proceedings of AAMAS 2008, pp. 283–290 (2008)
24. Taylor, M.E., Stone, P.: Transfer learning for reinforcement learning domains: a survey. J. Mach. Learn. Res. **10**, 1633–1685 (2009)
25. Taylor, M.E., Whiteson, S., Stone, P.: Transfer via inter-task mappings in policy search reinforcement learning. In: Proceedings of AAMAS 2007, pp. 156–163 (2007)

The *GRL* System: Learning Board Game Rules with Piece-Move Interactions

Peter Gregory[1](✉), Henrique Coli Schumann[1], Yngvi Björnsson[2], and Stephan Schiffel[2]

[1] Digital Futures Institute, Teesside University, Middlesbrough, UK
p.gregory@tees.ac.uk
[2] School of Computer Science, Reykjavik University, Reykjavik, Iceland
{yngvi,stephans}@ru.is

Abstract. Many real-world systems can be represented as formal state transition systems. The modeling process, in other words the process of constructing these systems, is a time-consuming and error-prone activity. In order to counter these difficulties, efforts have been made in various communities to learn the models from input data. One learning approach is to learn models from example transition sequences. Learning state transition systems from example transition sequences is helpful in many situations. For example, where no formal description of a transition system already exists, or when wishing to translate between different formalisms.

In this work, we study the problem of learning formal models of the rules of board games, using as input only example sequences of the moves made in playing those games. Our work is distinguished from previous work in this area in that we learn the interactions between the pieces in the games. We supplement a previous game rule acquisition system by allowing pieces to be added and removed from the board during play, and using a planning domain model acquisition system to encode the relationships between the pieces that interact during a move.

1 Introduction

Over the last decade, or ever since the advent of the *General Game-Playing (GGP)* competition [7], research interest in general approaches to intelligent game playing has become increasingly mainstay. GGP systems autonomously learn how to skilfully play a wide variety of (simultaneous or alternating) turn-based games, given only a description of the game rules. Similarly, *General Video-Game (GVG)* systems learn strategies for playing various video games in real-time and non-turn-based settings.

In the above mentioned systems the domain model (i.e., rules) for the game at hand is sent to the game-playing agent at the beginning of each match, allowing legitimate play off the bat. For example, games in GGP are described in a language named *Game Description Language (GDL)* [12], which has axioms for describing the initial game state, the generation of legal moves and how they alter

© Springer International Publishing Switzerland 2016
T. Cazenave et al. (Eds.): CGW 2015/GIGA 2015, CCIS 614, pp. 130–148, 2016.
DOI: 10.1007/978-3-319-39402-2_10

the game state, and how to detect and score terminal positions. Respectively, video games are described in the *Video Game Description Language (VGDL)* [18]. The agent then gradually learns improved strategies for playing the game at a competitive level, typically by playing against itself or other agents. However, ideally one would like to build fully autonomous game-playing systems, that is, systems capable of learning not only the necessary game-playing strategies but also the underlying domain model. Such systems would learn skilful play simply by observing others play.

Automated model acquisition is an active research area spanning many domains, including constraint programming and computer security (e.g. [1,2, 16]). There has been some recent work in GGP in that direction using a simplified subset of board games, henceforth referred to as *Simplified Game Rule Learner (SGRL)* [3]. In the related field of autonomous planning, the *LOCM* family of domain model acquisition systems [4–6] learn planning domain models from collections of plans. In comparison to other systems of the same type, these systems require only a minimal amount of information in order to form hypotheses: they only require plan traces, where other systems require state information.

In this work we extend current work on learning formal models of the rules of (simplified) board games, using as input only example sequences of the moves made in playing those games. More specifically, we extend the previous *SGRL* game rule acquisition system by allowing pieces to be added and removed from the board during play, and by using the *LOCM* planning domain model acquisition system for encoding and learning the relationships between the pieces that interact during a move, allowing modeling of moves that have side effects (such as castling in chess). Our work is thus distinguished from previous work in this area in that we learn the interactions between the pieces in the games.

The paper is structured as follows: the next section provides necessary background material on *LOCM* and *SGRL*, followed by a description of the combined approach. After this, a system for capturing game is described. This is followed by empirical evaluation and overview of related work, before concluding and discussing future work.

2 Background

In this section we provide background information about the LOCM and SGRL model acquisition systems that we base the present work on.

2.1 LOCM

The *LOCM* family of domain model acquisition systems [5,6] are inductive reasoning systems that learn planning domain models from only action traces. This is large restriction, as other similar systems require extra information (such as predicate definitions, initial and goal states, etc.). *LOCM* is able to recover domain information from such a limited amount of input due to assumptions about the structure of the output domain.

A full discussion of the *LOCM* algorithm is omitted and the interested reader is referred to the background literature [5,6,8] for more information. However, we discuss those aspects of the system as relevant to this work. We use the well-known Blocksworld domain as an example to demonstrate the form of input to, and output gained from, *LOCM*. Although a simple domain, it is useful as it demonstrates a range of features from both *LOCM* and *LOCM2* that are relevant to this work.

The input to the *LOCM* system is a collection of plan traces. Suppose, in the Blocks-world domain, we had the problem of reversing a two block tower, where block A is initially placed on block B. The following plan trace is a valid plan for this problem in the Blocksworld domain:

```
(unstack A B)
(put-down A)
(pick-up B)
(stack B A)
```

Each action comprises a set of indexed object transitions. For example, the unstack action comprises a transition for block A (which we denote as *unstack.1*) and another for block B (which we denote as *unstack.2*). A key assumption in the *LOCM* algorithm is that the behavior of each type of object can be encoded in one or more DFAs, where each transition appears at most once. This assumption means that for two object plan traces with the same prefix, the next transition for that object must exit from the same state. Consider plan trace 1 and 2 below:

```
1: (unstack A B)
1: (put-down A)

2: (unstack X Y)
2: (stack X Z)
```

In the first plan trace, block A is unstacked, before being put down on to the table. In the second plan trace, X is unstacked from block Y before being stacked on to another block, Z. The assumption that each transition only exists once within the DFA description of an object type means that the state that is achieved following an *unstack.1* transition is the same state that precedes both a *put-down.1* and a *stack.1* transition.

The output formalism of *LOCM* represents each object type as one or more parametrized DFAs. Figure 1 shows the output DFAs of *LOCM* for the Blocksworld domain. In this figure, we have manually annotated the state names in order to highlight the meanings of each state. The edges in the *LOCM* state machines represent object transitions, where each transition is labeled with an action name and a parameter position.

LOCM works in two stages: firstly to encode the structure of the DFAs, secondly to detect the state parameters.

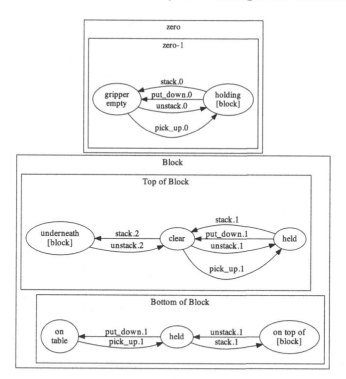

Fig. 1. State machines learned by *LOCM* in the Blocksworld planning domain. State labels are manually annotated to aid comprehension.

Blocks are the only type of object in Blocksworld, and are represented by the two state machines at the bottom of Fig. 1. Informally, these machines can be seen to represent what is happening above and below the block, respectively. In each planning state, a block is represented by two DFA states (for example, a block placed on the table with nothing above it would be represented by the 'clear' and 'on table' DFA states). Each of the block DFAs transition simultaneously when an action is performed, so when the previously discussed block is picked up from the table (transition pick_up.1) both the top and bottom machines move to the 'held' state.

The machine at the top of Fig. 1 is a special machine known as the zero machine (this refers to an imaginary zeroth parameter in every action) and this machine can be seen as encoding the structural regularities of the input action sequences. In the case of Blocksworld, the zero machine encodes the behavior of the gripper that picks up and puts down each block. Relationships between objects are represented by state parameters. As an example, in the 'Bottom of Block' machine, the state labelled 'on top of' has a block state parameter which represents the block directly underneath the current block.

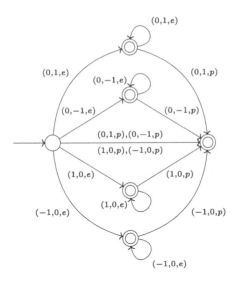

Fig. 2. A DFA, D_{rook}, describing the movements of a rook in chess

2.2   The *SGRL* System

The *SGRL* approach [3] models the movements of each piece type in the game
individually using a deterministic finite automata (DFA). One can think of the
DFA representing a language where each word describes a completed move (or
piece movement pattern) on the board and where each letter in the word —
represented by a triplet $(\Delta x, \Delta y, on)$— describes an atomic movement. The
triplet is read in the context of a game board position and a current square,
with the first two coordinates telling relative board displacement from the current square (file and rank, respectively) and the third coordinate telling the
content of the resulting relative square (the letter e indicates an empty square,
w an own piece, and p an opponent's piece). For example, $(0\ 1\ e)$ indicates
a piece moving one square up the board to an empty square, and the word
$(0\ 1\ e)\ (0\ 1\ e)\ (0\ 1\ p)$ a movement where a piece steps three squares up (over two
empty squares) and captures an opponent's piece. Figure 2 shows an example
DFA describing the movements of a rook, and Fig. 3 shows an example of how
moves on a chess and chess variant relate to micro-moves.

There are pros and cons with using the DFA formalism for representing
legitimate piece movements. One nice aspect of the approach is that well-known
methods can be used for the domain acquisition task, which is to infer from
observed piece movement patterns a consistent DFA for each piece type (we
are only concerned with piece movements here; for details about how terminal
conditions and other model aspects are acquired we refer to the original paper).
Another important aspect, especially for a game-playing program, is that state-
space manipulation is fast. For example, when generating legal moves for a piece
the DFA is traversed in a depth-first manner. On each transition the label of

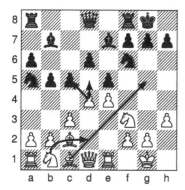 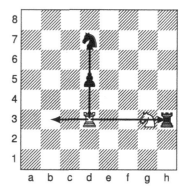

Fig. 3. A chess and a chess variant example. The left hand board shows an example of chess where potential moves are shown for the pawn on $d4$, advancing to $d5$ or capturing on $c5$. The former move yields the one-step piece-movement pattern $(0, 1, e)$ and the latter $(-1, 1, p)$. The knight move $b1$–$d2$ and the bishop move $c1$–$g5$ yield the piece-movement patterns $(2, 1, e)$ and $(1, 1, e)(1, 1, e)(1, 1, e)(1, 1, e)$, respectively. The *cannon* in Chinese chess slides orthogonally, but to capture it must leap over exactly one piece (either own or opponent's) before landing on the opponent's piece being captured. This is shown in the right hand board. Assuming the piece on $d3$ moves like a cannon, the move $d3$–$b3$ yields the piece-movement pattern $(-1, 0, e)(-1, 0, e)$, the move $d3$–$h3$ the pattern $(+1, 0, e)(+1, 0, e)(+1, 0, w)(+1, 0, p)$, and the move $d3$–$d7$ the pattern $(0, +1, e)(0, +1, p)(0, +1, e)(0, +1, p)$

an edge is used to find which square to reference and its expected content. If there are no matching edges the search backtracks. A transition into a final DFA state s generates a move in the form of a piece movement pattern consisting of the edge labels that were traversed from the start state to reach s. A special provision is taken to detect and avoid cyclic square reference in piece-movement patterns.

Unfortunately, simplifying compromises were necessary in *SGRL* to allow the convenient domain-learning mechanism and fast state-space manipulation. One such is that moves are not allowed to have side effects, that is, a piece movement is not allowed to affect other piece locations or types (with the only exception that a moving piece captures the piece it lands on). These restrictions for example disallow castling, en-passant, and promotion moves in chess. We now look at how these restrictions can be relaxed.

3 The *GRL* System

In this section, we introduce a board-game rule learning system that combines the strengths of both the *SGRL* and *LOCM* systems. One strength of the *SGRL* rule learning system is that it uses a relative coordinate system in order to generalize the input gameplay traces into concise DFAs. A strength of the *LOCM* system is that it can generalize relationships between objects that undergo simultaneous transitions.

One feature of the *SGRL* input language is that it represents the distinction between pieces belonging to the player and the opponent, but not the piece identities. This provides a trade-off between what it is possible to model in the DFAs and the efficiency of learning (far more games would have to be observed to learn if the same rules had to be learned for every piece that can be moved over or taken, and the size of the automata learned would be massive). In some cases, however, the identities of the interacting pieces are significant in the game.

We present the *GRL* system that enhances the *SGRL* system in such a way as to allow it to discover a restricted form of side-effects of actions. These side-effects are the rules that govern piece-taking, piece-production and composite moves. *SGRL* allows for limited piece-taking, but not the types such as in checkers where the piece taken is not on the destination square of the piece moving. Piece-production (for example, promotion in chess or adding men at the start of Nine Men's Morris) is not at all possible in *SGRL*. Composite moves, such as castling in chess or arrow firing in Amazons also cannot be represented in the *SGRL* DFA formalism.

3.1 The *GRL* Algorithm

The *GRL* algorithm can be specified in three steps:

1. Perform an extended *SGRL* analysis on the game traces (we call this extended analysis *SGRL+*). *SGRL* is extended by adding additional vocabulary to the input language that encodes the addition and removal of pieces from the board. Minor modifications have to be made to the consistency checks of *SGRL* in order to allow this additional vocabulary.
2. Transform the game traces and the *SGRL+* output into *LOCM* input plan traces, one for each game piece. Within this step, a single piece move includes everything that happens from the time a player picks up a piece until it is the next player's turn. For example, a compound move will have all of its steps represented in the plan trace.
3. Use the *LOCM* system to generate a planning domain model for each piece type. These domain models encode how a player's move can progress for each piece type. Crucially, unlike the *SGRL* automata, the learn domains refer to multiple piece types, and the relationships between objects through the transitions.

The output of this procedure will be a set of planning domains (one for each piece) which can generate the legal moves of a board game. By specifying a planning problem with the current board as the initial state, then enumerating the state space of the problem provides all possible moves for that piece. We now describe these three stages in more detail.

3.2 Extending *SGRL* Analysis for Piece Addition and Deletion

The first stage of *GRL* is to perform an extended version of the *SGRL* analysis. The input alphabet has been extended in order to include vocabulary to encode

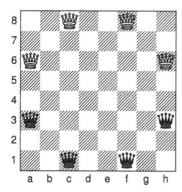 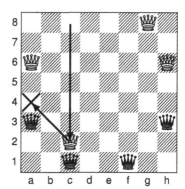

Fig. 4. The Game of Amazons is a two-player game typically played on a 10×10 board, shown here on an 8×8 board, in which each player has four pieces. The board on the left shows the initial setup of the game. On each turn, a player moves a piece as the queen moves in chess, before firing an arrow from the destination square, again in the move pattern of the queen in chess. The final square of the arrow is then removed from the game, thus after each move, one square is removed from the game. This can be seen on the right hand board: the queen moves from C8 to C2, before firing an arrow from C2 to A4 (A4 now cannot be transited). The loser is the first player unable to make a move.

piece addition and removal. To this end, we add the following two letters as possible commands to the DFA alphabet:

1. The letter 'a' which means that the piece has just been added at the specified cell (the relative coordinates of this move will always be $(0,0)$). The piece that is picked up now becomes the piece that is 'controlled' by the subsequent micro-moves.
2. The letter 'd' which means that the square has a piece underneath it which is removed from the game. This is to model taking pieces in games like peg solitaire or checkers.

The order in which these new words in the vocabulary are used is significant. An 'a' implies that the piece underneath the new piece remains in place after the move, unless it is on the final move of the sequence of micro-moves (this is consistent with the more limited piece-taking in *SGRL*, where pieces are taken only at the end of a move, if the controlled piece is another piece's square). For example, the following sequences both describe white pawn promotion in chess:

```
(0 1 e) (0 0 a) (0 0 d)
(0 1 e) (0 0 a)
```

Adding this additional vocabulary complicates the *SGRL* analysis in two ways: firstly in the definition of the game state, and secondly in the consistency checking of candidate DFAs. There is no issue when dealing with the removal of pieces and state definitions. There is, however, a small issue when dealing with piece addition. A move specified in *SGRL* input language is in the following form:

```
(12 (0 1 e) (0 0 a) (0 1 e))
```

This move specifies that the piece on square 12 moves vertically up a square, then a new piece appears (and becomes the controlled piece), which subsequently moves another square upwards. The issue is that for each distinct move, in *SGRL*, the controlled piece is defined by the square specified at the start of the move (in this case square 12). If a piece appears during a move, then *SGRL*+ needs to have some way of knowing which piece has appeared, in order to add this sequence to the prefix tree of the piece type. To mitigate this problem, we require all added pieces to be placed in the state before they are added, and as such they can now be readily identified. This approach allows us to learn the DFAs for all piece types, including piece addition.

The consistency algorithm of *SGRL* verifies whether a hypothesized DFA is consistent with the input data. One part of this algorithm that is not discussed in prior work is the *generateMoves* function, that generates the valid sentences (and hence the valid piece moves) of the state machine. There are two termination criteria for this function: firstly the dimensions of the board (a piece cannot move outside of the confines of the board), secondly on revisiting squares (this is important in case cycles are generated in the hypothesis generation phase). This second case is relaxed in *GRL* slightly since squares may be visited more than once due to pieces appearing and/or disappearing. However, we still wish to prevent looping behavior, and so instead of terminating when the same square is visited, we terminate when the same square is visited in the same *state* in the DFA.

With these two changes to *SGRL* analysis we have provided a means to representing pieces that appear and that remove other pieces from the board. However, the state machines do not tell us how the individual piece movement, additions and removals combine to create a complete move in the game. For this task, we employ use of the *LOCM* system, in order to learn the piece-move interactions.

3.3   Converting to *LOCM* Input

We use the *LOCM* system to discover relationships between the pieces and the squares, and to do this we need to convert the micro-moves generated by the *SGRL*+ DFAs to sequences of actions. To convert these, we introduce the following action templates (where X is the name of one of the pieces):

```
appearX ( square )
removeX ( square )
moveX ( squareFrom, squareTo )
moveOverX ( squareFrom, squareVia, squareTo )
moveAndRemoveX ( squareFrom, squareTo )
moveAndRemoveX ( squareFrom, squareVia, squareTo )
```

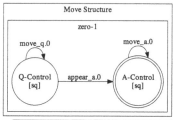

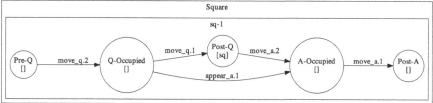

Fig. 5. The *LOCM* State Machines learned for the Queen piece in Amazons. The zero machine at the top of the figure shows the higher-level structure of the Queen's move. The Queen moves for a certain number of moves, then an arrow appears, and the arrow moves for a certain number of moves.

These action templates mirror the input language of the extended *SGRL+* system detailed above. Each plan that is provided as input to *LOCM* is a sequence of micro-moves that define an entire player's turn. When *LOCM* learns a model for each piece type, the interactions between the pieces and the squares they visit are encoded in the resultant domain model.

The input actions, therefore, encode the entire (possibly compound and with piece-taking side-effects) move. As an example from the Amazons (see Fig. 4) game, the following plan trace may be generated from the example micro-move sequence, meaning a queen is moved from A1 to B1, and then an arrow appears, and is fired to square C1:

```
Micro-move Sequence:
(0 (1 0 e) (0 0 a) (1 0 e))

Equivalent Plan Trace:
moveQueen ( A1, B1 )
appearArrow ( B1 )
moveArrow ( B1, C1 )
```

In all plan traces for the Queen piece in the Amazons game, the destination square of the Queen will always be the square in which the Arrow appears, and from which the Arrow piece begins its movement. Once the *LOCM* analysis is complete, this relationship is detected, and the induced planning domain will only allow arrows to be fired from the square that the queen moves to.

Figure 5 shows the actual *LOCM* state machines learned for the Queen piece from a collection of game traces. The top state machine represents the zero machine, which describes the general plan structure. The machine underneath represents one of the squares on the board. The parameters of the zero machine

```
(:action appear_a
 :parameters (?Sq1 - sq)
 :precondition
   (and (Q_Control ?Sq1)
        (Q_Occupied ?Sq1))
   :effect
   (and (A_Control ?Sq1)
        (not (Q_Control ?Sq1))
        (A_Occupied ?Sq1)
        (not (Q_Occupied ?Sq1)))))
```

Fig. 6. The PDDL Action for `appear_a`, representing the arrow appearing in the Amazons game.

represent the current square underneath the controlled piece. As can be seen from the generated PDDL action for *appear_a* (shown in Fig. 6), this parameter ensures that the arrow appears at the same location that the queen ended its move on. It also ensures that the moves are sequenced in the correct way: the Queen piece moves, the Arrow piece appears and finally the Arrow piece moves to its destination. The machine at the bottom of Fig. 5 encodes the transitions that a square on the board goes through during a Queen move. Each square undergoes a similar transition path, with two slightly different variants, and to understand these transitions it is important to see the two options of what happens on a square once a queen arrives there. The first option is that the Queen transits the square, moving on to the next square. The second option is that the Queen ends its movement; in this case, the Arrow appears, before transiting the square. These two cases are taken into account in the two different paths through the state machine.

As another example, consider the game of Checkers. The state machines generated by *LOCM* for the Pawn piece type are shown in Fig. 7. There are two zero machines in this example, which in this case means that *LOCM2* analysis was required to model the transition sequence, and there are separable behaviors in the move structure. These separable behaviors are the movement of the Pawn itself and of the King, if and when the Pawn is promoted. The top machine models the alternation between taking a piece and moving on to the next square for both Pawns and Kings, the bottom machine ensures that firstly a King cannot move until it has appeared, and secondly that a Pawn cannot continue to move once it has been promoted.

3.4   Move Generation

We now have enough information to generate possible moves based on a current state. The *GRL* automata can be used in order to generate moves, restricted by the *LOCM* state machines, where the *LOCM* machines define the order in which the pieces can move within a single turn.

The algorithm for generating moves is presented here as Algorithm 1. The algorithm takes as input an *SGRL+* state, a *LOCM* state and a square on the board. A depth-first search is then performed over the search space, in order to generate all valid traces. A state in this context is a triple of a board square, an *SGRL+* piece state and a *LOCM* state. Lines 2 to 3 define the termination criteria (a state has already been visited), lines 4 to 5 define when a valid move has been defined (when each of the *SGRL+* and *LOCM* DFAs are in a terminal state), and lines 6 to 7 define the recursion (in these lines, we refer to a state as being consistent. Consistency in this sense means that the state is generated by a valid synchronous transition in both the *SGRL+* and *LOCM* DFAs.

This completes the definition of the *GRL* system: an extended *SGRL* learns DFAs for each piece type, the *LOCM* system learns how these piece types combine to create entire moves, and then the move generation algorithm produces the valid moves in any given state. Next, we provide an evaluation of *GRL*, and detail the cost of each element of the system.

4   Game Trace Generation

One important aspect of learning game rules from traces is where our observations come from. The main results presented in this work are based on example traces generated from simulations of games generated from existing correct and complete rule sets. This is important because we can reliably evaluate the correctness of the learned rules if we have the true rules to test against. This is the main purpose of this work: to demonstrate the effectiveness of *GRL* in learning game rules, wherever the example traces come from.

However, to use the *GRL* system in the real world, we need to produce game traces from observations of real games. In this section we describe a visual tool for generating game traces. This tool allows the generic setup of a board game, and then allows the user to play out games in order to build a collection of game traces. The creation of a set of rules follows the following pattern:

1. **Setting the game up.** This consists of providing a name for the game, and specifying the number of piece types that are involved in the game per player and the dimensions of the board. For example, checkers is played on an 8×8 board, with two piece types (pawns and queens). Following this stage, a graphical view of the board is generated with a collection of piece prototypes to interact with for the next stage.
2. **Constructing the initial state of the game.** This stage requires the user to define the initial configuration of pieces graphically, by dragging the

---

Algorithm 1. Function to generate legal moves for a player from a combination of *SGRL+* and *LOCM* DFAs.

1: **function** GenerateMoves(sq,S_L,S_P)
2:    **if** Visited $\langle sq, S_L, S_P \rangle$ **then**
3:        **return**
4:    **if** terminal(S_L) and terminal(S_P) **then**
5:        add the current micro-move path to moves.
6:    **for all** consistent next states $\langle sq', S'_L, S'_P \rangle$ **do**
7:        GenerateMoves(sq', S'_L,S'_P)

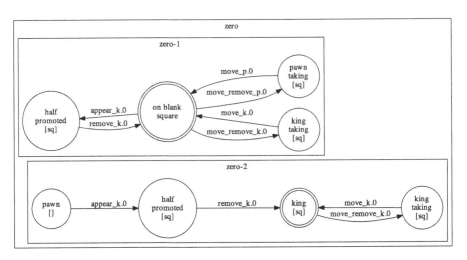

Fig. 7. The *LOCM* State Machines learned for the Pawn piece in the game of Checkers. The state machine incorporates the possibility of the Pawn being promoted into a King, and then subsequently can take other pieces.

piece prototypes to the square(s) that they start from. This completes the configuration of the board: the next stage is actually playing the game.

3. **Playing the initial game** (Fig. 8). The first game is played as a two player game, with the user playing both white and black moves. Within each turn, the player can move as many pieces as he or she chooses, add pieces to the board, and remove pieces from the board. These movements are interpreted as micro-moves, and after the user signals that the move has ended, these micro-moves become the representation for that 'turn' in the game. The exact way in which these micro-moves are calculated is explained below. After the user signals that the game is over, and which player (if any) has won the game, the *GRL* algorithm is used to learn an initial approximation of the game rules. This initial model learning seeds the next stage:

4. **Model refinement training.** Once an approximation on a set of rules exists, the user takes turns to play games for white and black pieces. The computer will make random moves for the opposing color during play. At the end of each game, all previous game traces are combined with the current game trace to learn a refined set of rules using the *GRL* system. After each game, the moves that the user has made can expand upon those in the current model. In the next game, the roles are reversed, with the computer controlling the opposite color than previously. Thus, with each game, the user both refines the moves for one of the colors and also observes informally whether or not the model appears to capture the true rules of the game. This stage is repeated until the user feels that the rules learned appear sufficiently close to the true game rules.

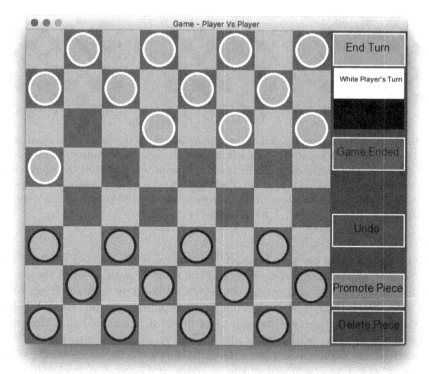

Fig. 8. Visual Game Trace Generator. This figure shows the gameplay screen for the game of checkers. White pieces are those with white borders, the piece type is denoted by the color in the center of the piece. Play proceeds by dragging the pieces, and the end of each turn and game is signaled by the options on the right hand side of the frame. (Color figure online)

4.1   Computing Micro-Move Representations

During each turn that a user makes, he or she may move pieces on the board, add pieces to the board and remove pieces from the board. These moves are broken down into micro-moves, in order to provide input to the *GRL* algorithm. All of these operations are performed using a mouse-driven point and click interface. Game pieces can be dragged using the mouse into their new positions. These new positions may not be adjacent to the square that the piece was lifted from, and so the question of how we formulate the micro-moves is important. One option would be to simply take the absolute move and encode this (left pane of Fig. 9). However, this has the consequence that important constraints on moves can be missed, as it appears that the piece can jump over whatever was in the way of it. Another option would be to track the mouse cursor as it dragged the piece to its destination, making each newly visited square part of the move (center pane of Fig. 9). This idea places a high burden on the user to move the mouse

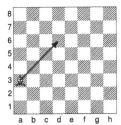 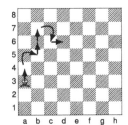 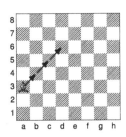

Fig. 9. Three different ways of generating micro-moves for the same piece move from visual mouse-dragged input. On the left, the exact move is encoded in one micro-move (3,3,e). The center option is to track the mouse movement of the user, which gives the messier (0,1,e),(1,1,e),(0,1,e),(0,1,e),(1,-1,e),(1,0,e). The option on the right, which we use, is to visit all squares on the shortest path between the start and the destination square: (1,1,e),(1,1,e),(1,1,e).

in an efficient way, in order to preserve the particular micro-move pattern. The method that we actually employ is to find the shortest path between the source and destination squares and encode this as the micro-move sequence (right hand pane of Fig. 9). If there is more than one shortest path, we encode all such paths.

When a piece is taken following a piece move, it is assumed that this piece transited the location of the taken piece. Thus we find the shortest path from the start to the taken piece's location and then on to the destination. When a piece is added to the board during a move, it begins a new element of a compound move, as described in Sect. 3. Taking these three elements (moving, adding and removing pieces) together, we have a method for computing entire micro-move sequences that then are accumulated to construct each game trace. We present this system in order to show how the *GRL* system can be used on concrete observations of game play. In future, we also intend to learn terminal states such that the computer attempts to beat the user, rather than simply making random moves. This may also give an indication of when learning has been effective enough, as a good guide to this could be when the computer player can beat the user.

5 Empirical Evaluation

In this section, we provide an evaluation of the *GRL* system. In order to perform this evaluation, we learn the rules of Amazons, Peg Solitaire and Checkers. We also provide evaluation for the three games Breakthrough, Breakthrough Checkers, and Breakthrough Chess as used in the *SGRL* evaluation [3] in order to demonstrate that performance is not adversely affected due to the changes in *GRL*, in the alphabet and consistency algorithm. We report on the time taken in the extended *SGRL+* phase, and the *LOCM* phase, to show the balance of time taken in each phase. We generate two forms of game traces for each game: one has a single move per turn (for these we generate 1000 game traces) and the

Table 1. Learning time in seconds to learn models for each of the pieces in the problem set with all moves known. (Using all moves only affects the time for the *SGRL+* element, as *LOCM* does not accept input when all moves are known).

Game	Piece	Element	Player1	Player2
Amazons	Queen	*SGRL+*	8.1	35.2
		LOCM	20.9	30.0
	Arrow	*SGRL+*	9.3	10.8
		LOCM	–	–
Checkers	Pawn	*SGRL+*	<0.01	<0.01
		LOCM	260.4	275.3
	King	*SGRL+*	<0.01	<0.01
		LOCM	52.9	47.0
Peg Solitaire	Peg	*SGRL+*	<0.01	–
		LOCM	8.8	–
BT	Pawn	*SGRL+*	<0.01	<0.01
CheckerBT	Checker	*SGRL+*	0.16	0.17
ChessBT	Pawn	*SGRL+*	<0.01	<0.01
	King	*SGRL+*	<0.01	<0.01
	Knight	*SGRL+*	<0.01	<0.01
	Bishop	*SGRL+*	3.3	3.4
	Rook	*SGRL+*	3.9	3.8
	Queen	*SGRL+*	12.3	12.5

other enumerates all possible moves (for these we generate 50 game traces). This method of evaluation is consistent with that chosen in the prior evaluation of *GRL*. All experiments are run on Mac OSX 10.10, running on an Intel i7 4650U CPU, with 8GB system RAM.

Table 1 shows the results of *GRL* on our benchmark problems. We show the time taken for each element of *GRL* to learn its model. We report both the time taken to learn the model for the first and second players (reported under 'Player1' and 'Player2'). In the case of Peg Solitaire, there is only one player, which explains the missing data. The missing data in the *LOCM* element for Amazons A piece is because an arrow piece in Amazons never starts a move itself and hence has no plan trace data. We first observe that the Breakthrough, Breakthrough Checkers and Breakthrough Chess results are not significantly different to the results for *SGRL*. Therefore the modifications made to the *SGRL* algorithm are not adversely affecting performance. We do not report *LOCM* time here, since these games do not require the *LOCM* analysis, as *SGRL+* is sufficient to represent them. It is notable that the time taken to learn the *LOCM* models is significantly larger than the time to learn the *SGRL+* individual piece models. The main cause of this difference is that the input plan traces are for

Table 2. Learning time in seconds to learn models for each of the pieces in the problem set with only a single move known per state. Results marked with a (*) returned an incorrect DFA.

Game	Piece	Element	Player1	Player2
Amazons	Queen	$SGRL+$	51.2*	54.9
		$LOCM$	20.9	30.0
	Arrow	$SGRL+$	86.2*	34.6*
		$LOCM$	–	–
Checkers	Pawn	$SGRL+$	<0.01	<0.01
		$LOCM$	260.4	275.3
	King	$SGRL+$	<0.01	<0.01
		$LOCM$	52.9	47.0
Peg Solitaire	Peg	$SGRL+$	<0.01	<0.01
		$LOCM$	8.8	–
BT	Pawn	$SGRL+$	<0.01	<0.01
CheckerBT	Checker	$SGRL+$	1.2*	3.1*
ChessBT	Pawn	$SGRL+$	<0.01	<0.01
	King	$SGRL+$	<0.01	<0.01
	Knight	$SGRL+$	<0.01	<0.01
	Bishop	$SGRL+$	43.2	41.0
	Rook	$SGRL+$	52.4	50.0
	Queen	$SGRL+$	145.3	140.4

every *turn* of the game, rather than the entire game trace. Because of this, $LOCM$ has 25,000 input traces for the Queen piece in Amazons and 11,000 for the Pawn piece in Checkers, for example. The time required to parse and analyze this number of plans is necessarily time consuming.

Table 2 shows the results of learning when only a single move per turn is known. Learning times are increased, as it takes longer to find consistent DFAs in the $SGRL+$ phase. Note that on several occasions, $SGRL+$ returns incorrect solutions. In these cases, there is simply insufficient input data. GRL is still effective in the majority of cases, however, and does not degrade the performance of $SGRL$ on the previous game data.

6 Related Work

Within the planning literature there are several domain model acquisition systems, each with varying levels of detail in their input observations. The *Opmaker2* system [13,17] learns models in the target language of OCL [14] and requires a partial domain model, along with example plans as input. The ARMS system [19], can learn STRIPS domain models with partial or no observation of intermediate states in the plans, but does at least require predicates to be declared. The LAMP system [20] can target PDDL representations with quantifiers and logical implications.

As for domain model acquisition in board games, an ILP approach exists for inducing chess variant rules from a set of positive and negative examples using background knowledge and theory revision [15]. Furthermore, [10] presents a system that learns games such as Connect4 and Breakthrough from video demonstrations using minimal background knowledge. Rosie [11] is an agent implemented in Soar that learns game rules (and other concepts) for simple board games via restricted interactions with a human. As for general video-game playing, a neuro-evolution algorithm showed good promise playing a large set of Atari 2600 games using little background knowledge [9].

7   Conclusions and Future Work

In this work, we presented a system capable of inducing the rules of more complex games than the current state-of-the-art system. This can help in both constructing the rules of games, or replacing the current move generation routine of an existing general game playing system (for fitting games). *GRL* improves *SGRL* by allowing piece addition, removal and structured compound moves. This was achieved by combining two techniques for domain-model acquisition, one rooted in game playing and the other in autonomous planning.

However, there remain interesting classes of board game rules that cannot be learned by *GRL*. One interesting rule class is that of a state-bound movement restriction. Games that exhibit this behavior allow only a subset of moves to occur in certain contexts: examples of this are check in chess (where only the subset of moves that exit check are allowed) and the compulsion to take pieces in certain varieties of checkers (thus restricting the possible moves to the subset that take pieces when forced). An approach to learning these restrictions could be developed given knowledge of all possible moves at each game state in the game.

Learning the termination criteria of a game is also an important step, if the complete set of rules of a board game is to be learned. This requires learning properties of individual states, rather than state transition systems. However, many games have termination criteria of the type that only one player (or no players) can make a move. For this class of game, and others based on which moves are possible, it should be possible to extend the *GRL* system to learn how to detect terminal states.

References

1. Aarts, F., De Ruiter, J., Poll, E.: Formal models of bank cards for free. In: 2013 IEEE Sixth International Conference on Software Testing, Verification and Validation Workshops, pp. 461–468. IEEE (2013)

2. Bessiere, C., Coletta, R., Daoudi, A., Lazaar, N., Mechqrane, Y., Bouyakhf, E.H.: Boosting constraint acquisition via generalization queries. In: ECAI, pp. 99–104 (2014)
3. Björnsson, Y.: Learning rules of simplified boardgames by observing. In: ECAI, pp. 175–180 (2012)
4. Cresswell, S., McCluskey, T., West, M.: Acquiring planning domain models using LOCM. Knowl. Eng. Rev. 28(2), 195–213 (2013)
5. Cresswell, S., Gregory, P.: Generalised domain model acquisition from action traces. In: International Conference on Automated Planning and Scheduling, pp. 42–49 (2011)
6. Cresswell, S., McCluskey, T.L., West, M.M.: Acquisition of object-centred domain models from planning examples. In: Gerevini, A., Howe, A.E., Cesta, A., Refanidis, I. (eds.) ICAPS. AAAI (2009)
7. Genesereth, M.R., Love, N., Pell, B.: General game playing: overview of the AAAI competition. AI Mag. 26(2), 62–72 (2005)
8. Gregory, P., Cresswell, S.: Domain model acquisition in the presence of static relations in the LOP system. In: International Conference on Automated Planning and Scheduling, pp. 97–105 (2015)
9. Hausknecht, M.J., Lehman, J., Miikkulainen, R., Stone, P.: A neuroevolution approach to general atari game playing. IEEE Trans. Comput. Intell. AI Games 6(4), 355–366 (2014)
10. Kaiser, L.: Learning games from videos guided by descriptive complexity. In: Hoffmann, J., Selman, B. (eds.) Proceedings of the Twenty-Sixth AAAI Conference on Artificial Intelligence, Toronto, Ontario, Canada, 22–26 July 2012, pp. 963–970. AAAI Press (2012)
11. Kirk, J.R., Laird, J.: Interactive task learning for simple games. In: Advances in Cognitive Systems, pp. 11–28. AAAI Press (2013)
12. Love, N., Hinrichs, T., Genesereth, M.: General game playing: game description language specification. Technical report, Stanford University, 4 April 2006. http://games.stanford.edu/
13. McCluskey, T.L., Cresswell, S.N., Richardson, N.E., West, M.M.: Automated acquisition of action knowledge. In: International Conference on Agents and Artificial Intelligence (ICAART), pp. 93–100 (2009)
14. McCluskey, T.L., Porteous, J.: Engineering and compiling planning domain models to promote validity and efficiency. Artif. Intell. 95(1), 1–65 (1997)
15. Muggleton, S., Paes, A., Santos Costa, V., Zaverucha, G.: Chess revision: acquiring the rules of chess variants through FOL theory revision from examples. In: De Raedt, L. (ed.) ILP 2009. LNCS, vol. 5989, pp. 123–130. Springer, Heidelberg (2010)
16. O'Sullivan, B.: Automated modelling and solving in constraint programming. In: AAAI, pp. 1493–1497 (2010)
17. Richardson, N.E.: An operator induction tool supporting knowledge engineering in planning. Ph.D. thesis, School of Computing and Engineering, University of Huddersfield, UK (2008)
18. Schaul, T.: A video game description language for model-based or interactive learning. In: Proceedings of the IEEE Conference on Computational Intelligence in Games (CIG 2013), pp. 193–200. IEEE (2013)
19. Wu, K., Yang, Q., Jiang, Y.: ARMS: an automatic knowledge engineering tool for learning action models for AI planning. Knowl. Eng. Rev. 22(2), 135–152 (2007)
20. Zhuo, H.H., Yang, Q., Hu, D.H., Li, L.: Learning complex action models with quantifiers and logical implications. Artif. Intell. 174, 1540–1569 (2010)

Creating Action Heuristics for General Game Playing Agents

Michal Trutman[1] and Stephan Schiffel[2]([⊠])

[1] Faculty of Information Technology,
Brno University of Technology, Brno, Czech Republic
`mtrutman@gmail.com`
[2] School of Computer Science, Reykjavík University, Reykjavík, Iceland
`stephans@ru.is`

Abstract. Monte-Carlo Tree Search (MCTS) is the most popular search algorithm used in General Game Playing (GGP) nowadays mainly because of its ability to perform well in the absence of domain knowledge. Several approaches have been proposed to add heuristics to MCTS in order to guide the simulations. In GGP those approaches typically learn heuristics at runtime from the results of the simulations. Because of peculiarities of GGP, it is preferable that these heuristics evaluate actions rather than game positions. We propose an approach that generates heuristics that estimate the usefulness of actions by analyzing the game rules as opposed to the simulation results. We present results of experiments that show the potential of our approach.

Monte-Carlo Tree Search (MCTS) with UCT (Upper Confidence bounds applied to Trees) [11] has seen wide-spread success in recent years and is the state-of-the-art in General Game Playing (GGP) [9] today. One reason for the success of MCTS in GGP is that MCTS can find good moves even in the absence of domain knowledge in the form of evaluation functions or heuristics. However, that does not mean that MCTS cannot benefit from heuristics, if they are available. In fact, there are several examples, such as [5] or [21] in GGP and [13] in game specific settings that show how heuristics can be used in MCTS to improve the performance of a game player. In General Game Playing, a program is presented with the rules of a previously unknown game and needs to play this game well without human intervention. Thus, the main problem of using domain knowledge in the form of heuristics in a General Game Playing program is, that the program must generate or learn the heuristics automatically for the game at hand.

Heuristics used in search come traditionally in the form of a state evaluation function, that is, an evaluation of non-terminal states in a game. Especially in GGP, it seems advantageous to evaluate actions instead of states. In perfect-information turn-taking games, there is no difference between the value of an action in a given state and the value of the state that is reached by executing that action. However, games in GGP can have simultaneous moves in which case the successor state depends on the actions of all players. Even in the case of turn-taking games, evaluating an action directly instead of the successor state

© Springer International Publishing Switzerland 2016
T. Cazenave et al. (Eds.): CGW 2015/GIGA 2015, CCIS 614, pp. 149–164, 2016.
DOI: 10.1007/978-3-319-39402-2_11

reached by that action saves the time needed to compute one state update. In GGP, this time is often significant [18], unless dominated by the time for computing the heuristics. Both reasons make it beneficial to evaluate actions instead of states, especially in the context of MCTS. Previous approaches to generate action heuristics in GGP are very limited in the sense that the learned heuristics are very simple and often ignore the context (state) in which an action is executed.

In the current work, we are exploring whether more accurate action heuristics can be generated by analyzing the rules of a game instead of learning them from simulation results. The paper is organized as follows: In the next section we give a brief background of the game description language and MCTS. Then we introduce our approach to generate action heuristics and, finally, we evaluate the approach and discuss the results.

1   Preliminaries

1.1   Game Description Language

Games in GGP are described in the so-called *Game Description Language* (GDL) [14]. A game description in GDL is a logic program with a number of predefined predicates and restrictions to ensure finiteness of derivations. The rules in the program can be used to compute an initial state, legal moves, successor states, terminality of states and goal values of the players. Thus, they are sufficient to simulate the game. GDL permits to describe a large range of deterministic perfect-information simultaneous-move games with an arbitrary number of adversaries. Turn-based games can be modeled by only allowing a move with no effect for players that do not have a turn (a *noop* move). Predefined predicates have a game-specific semantic, such as for describing the initial game state (*init*), detecting (*terminal*) and scoring (*goal*) terminal states, and for generating (*legal*) moves and successor states (*next*). Each game state can be represented by the set of facts that hold in the state (e.g., *cell*(1, 1, b)).

The following figure shows a partial GDL description for a variant of the game Tic Tac Toe, where the goal was reduced to build any of the two diagonal lines on the board.

```
1 role(xplayer). role(oplayer).
2 init(cell(1, 1, b)) ... init(cell(3, 3, b)).
3 init(control(xplayer)).
4 legal(W, mark(X, Y)) :-
5     true(cell(X, Y, b)), true(control(W)).
6 legal(oplayer, noop) :- true(control(xplayer)).
7 ...
8 next(cell(M, N, x)) :-
9     does(xplayer, mark(M, N)), true(cell(M, N, b)).
10 next(control(oplayer)) :- true(control(xplayer)).
11 ...
12 diagonal(X) :- true(cell(1, 1, X)),
```

```
13      true(cell(2, 2, X)), true(cell(3, 3, X)).
14 goal(xplayer, 100) :- diagonal(x).
15 goal(xplayer, 0) :- diagonal(o).
16 terminal :- diagonal(x).
17 ...
```

1.2   Monte-Carlo Tree Search

Monte-Carlo Tree Search works by running complete simulations of a game, that is, repeatedly playing a simulation of a game starting at the current state and stopping when a terminal state is reached. The simulations are used to gradually build a game tree in memory. The nodes in this tree store the average reward (goal value) achieved by executing a certain action in a certain state. When the deliberation time is up, the player plays the best move in the root node of the tree. Each simulation consists of four steps:

1. *Selection:* selecting the actions in the tree based on their average reward until a leaf node of the tree is reached,
2. *Expansion:* adding one or several nodes to the tree,
3. *Playout:* playing randomly from the leaf node of the tree until a terminal state is reached,
4. *Back-Propagation:* updating the values of the nodes in the tree with the reward achieved in the playout.

2   Related Work

The most common modification of MCTS algorithm is MCTS with UCT [11] allowing to set a trade-off between exploration and exploitation. One of the first attempts to enrich MCTS/UCT with a heuristic was a *progressive bias* added to the UCT formula to direct search according to possibly expensive heuristic knowledge in Go [1].

There are two ways to create heuristics. First, offline heuristics rely on game analysis and feature detection before the game starts. Once the heuristic is generated, it is used throughout the game. On the other hand, online heuristics are learned and improved during game play.

Several ways were suggested of how to automatically generate heuristic offline. While [12] try to build a heuristic upon detecting common game features like a game board, game pieces or quantities; [22] look for more generic and game independent concepts. [2] uses game properties like termination, control over the board and payoff as components in his evaluation function. In [19] fuzzy logic is used to evaluate the goal condition in an arbitrary state and the value is used as a measure of how close the state is to a goal state. The approach is further improved by using feature discovery and was used in FLUXPLAYER, when winning the GGP competition in 2006.

A different approach relies on learning a heuristic online from simulations of the game. The first notable enhancement of MCTS was *Rapid Action Value Estimation* (RAVE) [8], a method to speed up the learning process of action values inside the game tree. A similar technique to learn state and move knowledge was based on which state fluents mostly occur in the winning states and which moves lie on the winning paths [20]. The state-of-the-art has also been greatly advanced by *Move-Average Sampling Technique* (MAST) [5]. MAST is a control scheme used in the playout phase of MCTS which learns the general value of an action independent from the context the action is used in. This and other control schemes such as *Features-to-Action Sampling Technique* (FAST) [7], early cutoffs and unexplored action urgency [4] were used by CADIAPLAYER, a successful player that won the GGP competition three times. Recently, the MAST concept was made more accurate by using sequences of actions of given length (N-grams) instead of just single actions [21]. It has also been shown, that is possible to get more information from the playouts by assessing the lengths of simulations and evaluating the quality of the terminal state reached [16].

As it was shown in [3], MCTS converges slowly to the true Minimax value and therefore different combinations of Minimax and MCTS were suggested. While [3] use a different operator for backing up the values through the tree instead of just averaging them; [23] introduce MCTS Solver, an $\alpha\beta$-search-like approach to prove correct values of fully expanded parts of the game tree in Lines of Actions. Recently, [13] experiment with calculating approximate Minimax backups from heuristic values to further improve node selection in Kalah, Breakthrough and Lines of Action. However, the heuristic is currently built on game specific knowledge.

3   Generating Action Heuristic

Our idea for generation of an action heuristic is to create an action-based version of the state evaluation function described in [19], which uses fuzzy logic to evaluate the degree of truth of a goal condition. Turning it into an action heuristic is achieved by taking the goal condition, regressing it one step and filtering it according to an action a of a player p. This yields a new condition which – when satisfied in the current game state – allows player p to achieve the goal condition by executing action a.

3.1   Regression

Our definition of regression is based on regression in the *situation calculus* as defined in [17]. Similar to situation formulas in situation calculus, we define a *state formula* in a game as any first-order formula over the predicate, function and constant symbols of the game description with the exception of the does predicate and any predicate depending on does.

Thus, the truth value of a state formula can be determined in any state independently on the actions that players choose in that state. For example, goal($xplayer, 100$) is a state formula in our Tic Tac Toe game, because `goal` may not depend on `does` according to GDL restrictions. For the purpose of this paper, we only consider variable-free state formulas and game descriptions. Generalizing the proposed algorithm to non-ground game descriptions should be straightforward.

The regression of a variable-free state formula F by one step, denoted as $\mathcal{R}[F]$, is defined recursively as follows:

- $\mathcal{R}[\text{true}(X)] = F_1 \vee F_2 \vee \ldots \vee F_n$
  where F_i are the bodies of all rules of the following form in the game description: $\text{next}(X) :- F_i$
- $\mathcal{R}[\text{distinct}(a_1, a_2)] = \text{distinct}(a_1, a_2)$
- If F is any atom $p(a_1, a_2, \ldots, a_n)$ other than true or distinct, then

$$\mathcal{R}[F] = \mathcal{R}[F_1] \vee \mathcal{R}[F_2] \vee \ldots \mathcal{R}[F_n]$$

  where F_i are the bodies of all rules with head $p(a_1, a_2, \ldots, a_n)$.
- If F is a non-atomic formula then the regression is defined as follows:

$$\mathcal{R}[F_1 \vee F_2] = \mathcal{R}[F_1] \vee \mathcal{R}[F_2]$$
$$\mathcal{R}[F_1 \wedge F_2] = \mathcal{R}[F_1] \wedge \mathcal{R}[F_2]$$
$$\mathcal{R}[\neg F] = \neg \mathcal{R}[F]$$

Note, that the regression of a state formula is not a state formula in general. On the contrary, replacing $\text{true}(X)$ with the $\text{next}(X)$ during the regression, introduces dependencies on the `does` predicate and thus the executed moves. These dependencies will be used in the following section to define a heuristic function for each action.

3.2   Algorithm

Based on the previous definition, we propose the following algorithm to generate a heuristic function for each action a of a player p. The algorithm consists of following steps:

1. Compute $\mathcal{R}[\text{goal}(p, 100)]$, the regression of $\text{goal}(p, 100)$.[1] $\mathcal{R}[\text{goal}(p, 100)]$ represents a condition on a state and actions of players that – when fulfilled – allow to reach a goal state for player p.
2. $\mathcal{R}[\text{goal}(p, 100)]$ contains conditions on actions of players. However, we want a formula that indicates when it is a good idea for player p to execute action a. To obtain such a formula, we restrict $\mathcal{R}[\text{goal}(p, 100)]$ to those parts that are consistent with $does(p, a)$. In practice this is achieved by replacing all occurrences of $does(r, b)$ for any r and b in $\mathcal{R}[\text{goal}(p, 100)]$ as follows:

---

[1] In the current implementation, we take into consideration only the highest valued goal for each player. Combining different goals could be done similar to the way described in [19], but would make the heuristics more expensive to compute.

(a) In case $p = r$ and $a = b$, the occurrence of does(r, b) is replaced with the boolean constant `true`.

(b) In case $p = r$, but $a \neq b$, the occurrence of does(r, b) is replaced with the boolean constant `false`.

(c) In case $p \neq r$, the condition is on an action for another player. In that case, the replacement depends on whether or not the game is turn-taking[2]. If the game has simultaneous moves, does(r, b) is replaced with the unknown value in three-valued logic. This represents, that we are not sure, which action the opponent decides to play. In case of a turn-taking game, if b is a noop action, the does(r, b) is replaced with `true`, otherwise with `false` (because r must do a noop action if p is doing a non-noop action such as a).

3. The formula is simplified according to laws of three-valued logic. In particular, any boolean values introduced by the previous replacement step are propagated up and the formula is partially evaluated.

3.3   Tic Tac Toe

Let us demonstrate, how the algorithm works on the simplified version of the game Tic Tac Toe, where the goal was reduced to build only some of the two diagonal lines. The grounded and expanded version of the goal for the player $xplayer$ is

$$\big(\text{true}(cell(1,1,x)) \wedge \text{true}(cell(2,2,x)) \wedge \text{true}(cell(3,3,x))\big) \vee$$
$$\vee \big(\text{true}(cell(1,3,x)) \wedge \text{true}(cell(2,2,x)) \wedge \text{true}(cell(3,1,x))\big) \qquad (1)$$

Assume, we are computing the heuristic function for the action $mark(1,1)$ for the role $xplayer$. The regression of the goal above will replace all occurrences of true(X) with the bodies of the respective next rules. For example, for true$(cell(1,1,x))$, the grounded game description contains following $next$ rules with matching arguments:

```
next(cell(1, 1, x)) :- true(cell(1, 1, b)),
    does(xplayer, mark(1, 1)).
next(cell(1, 1, x)) :- true(cell(1, 1, x)).
```

To regress $true(cell(1,1,x))$, we replace it with the disjunction of the bodies of the $next$ rules:

$$\big(\text{true}(cell(1,1,b)) \wedge \text{does}(xplayer, mark(1,1))\big) \vee \text{true}(cell(1,1,x)) \qquad (2)$$

does$(xplayer, mark(1,1))$ in (2) is further replaced with boolean true, because both role and action match the ones we are interested in right now.

$$\big(\text{true}(cell(1,1,b)) \wedge \text{T}\big) \vee \text{true}(cell(1,1,x)) \qquad (3)$$

---

[2] We detect whether a game is turn-taking and also which action is the noop action, using a theorem prover [10] or using random simulation in case theorem proving does not yield an answer. The cost of this is negligible compared to, e.g., grounding the game rules.

The formula (3) can be simplified, which yields (4) as the final replacement for true($cell(1, 1, x)$):

$$\text{true}(cell(1, 1, b)) \lor \text{true}(cell(1, 1, x)) \tag{4}$$

Going back to the goal condition (1), the next part of the formula to be regressed is true($cell(2, 2, x)$). The matching *next* rules are:

```
next(cell(2, 2, x)) :- true(cell(2, 2, b)),
   does(xplayer, mark(2, 2)).
next(cell(2, 2, x)) :- true(cell(2, 2, x)).
```

This time, the does($xplayer, mark(2, 2)$) is replaced with boolean false, because the role matches, but the action does not. This yields formula (5), which can be simplified to (6). This equals the term we have started with, meaning that true($cell(2, 2, x)$) stays in the formula untouched.

$$\big(\text{true}(cell(2, 2, b)) \land F\big) \lor \text{true}(cell(2, 2, x)) \tag{5}$$
$$\text{true}(cell(2, 2, x)) \tag{6}$$

We repeat the same steps for any other *true* keywords in the goal (1). As in the previous case, each term is replaced by the term itself and nothing in the formula is changed. The final heuristic formula for *xplayer* taking action $mark(1, 1)$ is:

$$\big(\text{true}(cell(2, 2, x)) \land \text{true}(cell(3, 3, x)) \land$$
$$\land (\text{true}(cell(1, 1, b)) \lor \text{true}(cell(1, 1, x)))\big) \lor$$
$$\lor \big(\text{true}(cell(3, 1, x)) \land \text{true}(cell(2, 2, x)) \land \text{true}(cell(1, 3, x))\big) \tag{7}$$

As can be seen, this condition describes a situation in which *xplayer* taking action $mark(1, 1)$ would lead to a winning state. Thus, a boolean evaluation of such conditions for all legal moves in a state is equivalent to doing 1-ply lookahead.

3.4   Evaluation

Using the algorithm above, the heuristic formula is constructed for any role and any potentially legal move during the start clock. During game play, the formula for each legal action is evaluated against the current game state s using fuzzy logic as described in [19], but with different t-norm and t-co-norm, as the original ones proved to be too slow. For non-atomic formulas the evaluation function is defined as

$$eval(f \land g, s) = \top(eval(f, s), eval(g, s))$$
$$eval(f \lor g, s) = \bot(eval(f, s), eval(g, s))$$
$$eval(\neg f, s) = 1 - eval(f, s)$$

where $\top$ and $\bot$ are the product t-norm and t-co-norm:

$$\top(a, b) = a \cdot b$$
$$\bot(a, b) = a + b - a \cdot b$$

All remaining atoms of the heuristic formula are of the form true(X), these are evaluated as

$$eval(true(f), s) = \begin{cases} p & \text{if } f \text{ is true in } s \\ 1 - p & \text{otherwise} \end{cases}$$

We used $p = 0.97$ for our experiments.

Thus, our action heuristics can be defined as $H(s, r, a) = eval(F_{r,a}, s)$, where $F_{r,a}$ is the heuristic formula constructed for role r and action a.

In essence, a higher value of the evaluation means, that more prerequisites are satisfied for a particular action to lead to a goal state.

The heuristic values for each legal action in the initial state of the aforementioned version of the game Tic Tac Toe are shown in the following table.

(1, 3)	(2, 3)	(3, 3)
0.0026	0.0018	0.0026
(1, 2)	(2, 2)	(3, 2)
0.0018	0.0035	0.0018
(1, 1)	(2, 1)	(3, 1)
0.0026	0.0018	0.0026

The heuristic function was evaluated in the initial game state (empty board). We can see, that the action with the highest value, is to take the middle cell ($mark(2, 2)$), followed by 4 actions taking one of the corner cells. Indeed, this corresponds with the fact that marking the middle cell as the first action leads to the most options for winning the game.

In [15, 19], we described methods for improving the fuzzy evaluation by using additional knowledge about the game for the evaluation of the atoms. The same methods can be used for the action heuristics presented here. For the experiments presented in the next section, we restricted ourselves to using only very limited additional knowledge especially selected for not increasing the time for evaluating the heuristics significantly. In particular, we only use knowledge about persistent fluents as defined in [10].

A fluent is *persistent true* if, once it holds in a state, it will persist to hold in all future states. For example, $cell(1, 1, x)$ is *persistent true* in Tic Tac Toe. A fluent is *persistent false* if, once it does not hold in a state, it will never hold in any future state. For example, $cell(1, 1, b)$ is *persistent false* in Tic Tac Toe.

Information like this can be detected in some, but not all of the games we tested. In case we could (automatically) infer this knowledge, we modify the evaluation function as follows:

$$eval(true(f), s) = \begin{cases} 1 & \text{if } f \text{ is true in } s \text{ and} \\ & f \text{ is persistent true} \\ p & \text{if } f \text{ is true in } s \text{ and} \\ & f \text{ is not persistent true} \\ 0 & \text{if } f \text{ is false in } s \text{ and} \\ & f \text{ is persistent false} \\ 1 - p & \text{otherwise} \end{cases}$$

4   Search Controls

In this section we recapitulate three ways, how an action heuristic can be utilized in the MCTS player to control different part of the search. All methods are described in [7]. While the first method uses the heuristic to guide the random playouts, in the second one it controls the search tree growth in the selection phase. The last approach presented is a combination these two concepts. We will use these methods later together with our own heuristics.

4.1   Playout Heuristic

In the standard MCTS with UCT, actions are selected uniformly at random during the playout phase. However, if we have any information on which actions are good, it is better to bias the action selection in favor of more promising moves [5]. This can be accomplished using the Gibbs (Boltzmann) distribution:

$$P(s, r, a) = \frac{e^{H(s,r,a)/\tau}}{\sum_{a'} e^{H(s,r,a')/\tau}}$$

where $P(s, r, a)$ is the probability that the action a will be chosen by role r in the current playout state s and $H(s, r, a)$ is the action heuristic function. The parameter τ is *temperature* and specifies how random actions are chosen. Whereas the high values makes it rather uniform; $\tau \to 0$ means that more valued actions are chosen more likely. Based on trial and error testing, good values for τ lie somewhere between 0.5 and 2. We used $\tau = 1$ in our tests.

One drawback of this method is, that the heuristic function must be evaluated for all legal moves in every playout state within the simulation, which is sometimes too costly.

4.2   Tree Heuristic

An action heuristic commonly used in the game tree is *Rapid Action Value Estimation* (RAVE) [8]. It keeps a special value $Q_{RAVE}(s, r, a)$, which is an average outcome of simulations, where action a was taken by role r in any state on the path below s.

In our case, instead of Q_{RAVE}, we use the value $H(s, r, a)$ provided by the action heuristic described earlier. Initially, only the heuristic is used to give an estimate of the action value, but as the sampled action value $Q(s, r, a)$ becomes more reliable with more simulations executed, it should be more trusted over the heuristic $H(s, r, a)$. This is achieved by using a weighted average as in RAVE:

$$Q(s, r, a)' := \beta(s) \times H(s, r, a) + (1 - \beta(s)) \times Q(s, r, a)$$

with

$$\beta(s) = \sqrt{\frac{k}{3 \times N(s) + k}}$$

The *equivalence parameter* k controls, how many simulations are needed for both estimates to have equal weights. The $N(s)$ function returns number of visits of the state s.

Before use, all heuristic values $H(s, r, a)$ are normalized and scaled into the range 0 to 100 which is the range of possible game outcomes. We use $k = 20$ in our tests, as it turned out to work best with the heuristic function we use.

4.3   Combined Heuristic

Both of the previous control schemes can be combined together and a heuristic can be used to guide the action selection both during the random playouts and in the game tree. As was shown in [6], this combination has a synergic effect, when they used RAVE as a tree heuristic and MAST as a playout heuristic. In our case, we use the action heuristic we developed in both control schemes with the same parameters as mentioned above.

5   Results

We run two sets of experiments. First, we matched players using the three aforementioned concepts of the action heuristic (playout, tree and combined heuristic) against a pure MCTS/UCT player with constant number of simulations per turn. Then, we matched playout, tree and combined heuristic against MAST, RAVE and MAST+RAVE players, respectively. We did not match MAST against our tree heuristic, because they operate in different stages of MCTS and the results are not comparable. Similarly for playout heuristics vs. RAVE. Constant time per turn was used in this experiment. We set temperature $\tau = 10$ for MAST and equivalence parameter $k = 1000$ for RAVE as recommended in [6].

Each set consists of 300 matches per game with each control scheme. The tests were run on Linux with multicore Intel(R) Xeon(R) 2.40 GHz processor with 4 GB memory limit and 1 CPU core assigned to each agent. Rules for all the games tested can be found in the game repository at games.ggp.org.

Table 1. Tournament using playout, tree and combined heuristic against pure MCTS player with fixed number of simulations.

Game	No heur. vs. Playout h.	No heur. vs. Tree h.	No heur. vs. Combined h.
battle	**90.2** × 80.9 (±2.4)	87.3 × **95.1** (±2.1)	**91.6** × 84.9 (±2.0)
bidding-tictactoe	19.2 × **80.8** (±3.5)	42.8 × **57.2** (±3.0)	19.7 × **80.3** (±3.6)
blocker	**61.0** × 39.0 (±5.5)	48.0 × 52.0 (±5.7)	**67.3** × 32.7 (±5.3)
breakthrough	51.3 × 48.7 (±5.7)	47.3 × 52.7 (±5.7)	48.7 × 51.3 (±5.7)
checkers-small	**94.3** × 5.7 (±2.4)	50.2 × 49.8 (±4.4)	**96.0** × 4.0 (±1.9)
chinesecheckers2	**81.0** × 69.0 (±2.8)	75.3 × 74.6 (±2.8)	**85.7** × 64.3 (±2.6)
chinook	**76.0** × 32.0 (±5.1)	54.7 × 56.3 (±5.6)	**76.0** × 37.0 (±5.2)
connect4	**81.2** × 18.8 (±4.1)	**56.8** × 43.2 (±5.4)	**86.7** × 13.3 (±3.6)
crisscross	**75.3** × 49.8 (±4.0)	62.3 × 62.8 (±4.2)	**74.3** × 50.8 (±4.0)
ghostmaze2p	28.7 × **71.3** (±3.1)	19.5 × **80.5** (±3.3)	30.8 × **69.2** (±3.4)
9BoardTicTacToe	26.7 × **73.3** (±5.0)	32.3 × **67.7** (±5.3)	22.7 × **77.3** (±4.7)
pentago_2008	22.7 × **77.3** (±4.5)	35.5 × **64.5** (±5.1)	21.2 × **78.8** (±4.3)
sheep_and_wolf	**74.7** × 25.3 (±4.9)	51.7 × 48.3 (±5.7)	**78.7** × 21.3 (±4.6)
skirmish	70.1 × 69.0 (±1.1)	79.4 × 77.3 (±1.5)	**78.4** × 75.0 (±1.5)

5.1 Playing Strength

Table 1 shows the average scores reached by the pure MCTS and the heuristic agent along with a 95 % confidence interval. We allowed 10000 simulations per turn and the time spent on evaluating the heuristic was measured.

The game with the strongest position of the combined scheme is Bidding Tic Tac Toe with a score 80 (±3.5) against 20; it is also good in Nine Board Tic Tac Toe, Pentago and Ghost Maze. On the other hand, it is particularly bad in Checkers and Connect4, but closer look reveals, that this is only because of the playout heuristic, while the tree heuristic has not much influence in these games. The playout heuristic follows the similar trend as the combined scheme. The tree heuristic is also significantly better in Battle and there is no game with totally hopeless result as there was with the playout heuristic.

It is also worth mentioning, how the results in the combined control scheme are connected to the playout and the tree heuristic. It seems, the influence of the playout heuristic on the overall result is much higher. Especially, when it is useless or even misleading, then the combined result is dragged down by it (Checkers, Connect4). It seems that the playout heuristic is more vulnerable, while the tree heuristic control scheme can recover when the heuristic is misleading. Thus, it is essential for the playout heuristic to be good, if it is used.

Bidding Tic Tac Toe is a game, where two players are bidding coins in order to mark a cell with an objective to build a line of three symbols as in Tic Tac Toe. However, the key property of this game is the bidding part – by doing

Table 2. Tournament against MAST and RAVE player with constant time per turn.

Game	MAST vs. Playout h.	RAVE vs. Tree h.	MAST+RAVE vs. Combined h.
battle	**95.0** × 83.7 (±1.8)	87.9 × **96.1** (±2.0)	**94.1** × 80.1 (±1.8)
bidding-tictactoe	26.2 × **73.8** (±4.2)	41.1 × **58.9** (±3.1)	26.5 × **73.5** (±4.4)
blocker	**60.0** × 40.0 (±5.5)	50.8 × 49.2 (±5.7)	**59.3** × 40.7 (±5.6)
breakthrough	45.7 × 54.3 (±5.6)	50.0 × 50.0 (±5.7)	52.3 × 47.7 (±5.7)
checkers-small	**95.0** × 5.0 (±2.1)	51.0 × 49.0 (±4.3)	**95.0** × 5.0 (±2.0)
chinesecheckers2	**96.4** × 53.3 (±1.8)	74.8 × 75.2 (±2.8)	**95.1** × 54.5 (±1.7)
chinook	**84.0** × 30.7 (±4.7)	52.0 × 59.0 (±5.6)	**86.3** × 28.0 (±4.5)
connect4	**83.3** × 16.7 (±4.0)	52.8 × 47.2 (±5.5)	**83.0** × 17.0 (±3.9)
crisscross	62.8 × 62.3 (±4.24)	62.3 × 62.8 (±4.2)	62.5 × 62.5 (±4.2)
ghostmaze2p	31.3 × **68.7** (±3.4)	20.2 × **79.8** (±3.2)	33.3 × **66.7** (±3.5)
9BoardTicTacToe	34.3 × **65.7** (±5.4)	35.3 × **64.7** (±5.4)	31.3 × **68.7** (±5.3)
pentago_2008	21.2 × **78.8** (±4.4)	42.3 × **57.7** (±5.3)	20.3 × **79.7** (±4.2)
sheep_and_wolf	**77.0** × 23.0 (±4.8)	53.7 × 46.3 (±5.6)	**76.3** × 23.7 (±4.8)
skirmish	80.5 × 79.7 (±1.5)	79.7 × 77.6 (±1.5)	**84.8** × 75.3 (±1.)

wrong bids, the game can be easily lost, even when the markers are placed in good positions on the board. The action heuristic does not help in any way with the bidding, it only helps to arrange the markers in a line. In spite of this, all the heuristic agents still have a significant advantage in this game. One explanation is that when a player wins a bid, it can actually use its move well which makes especially the playouts more reliable.

Table 2 shows results for matches played against MAST, RAVE and MAST+ RAVE. Although MAST and MAST+RAVE outperform the playout and the combined heuristic in most of the games, they still hold their good position in Pentago or Nine Board Tic Tac Toe. The heuristic performs surprisingly well against RAVE with no significant loss. By comparing Tables 1 and 2, we see that our action heuristics perform well against MAST and RAVE in exactly the same games in which they did well against a pure MCTS/UCT player. This suggests, that our approach is complementing MAST and RAVE by improving performance in games in which MAST and RAVE do not seem to have much positive effect.

In general, it can be said, that the heuristic player performs very well in Tic Tac Toe-like games, as they contain many persistently true fluents and the heuristic is built in such a way, that it leads the player to the goal directly.

Table 3. Percentage of the game time spent on evaluating the action heuristic and relative number of simulations compared to the non-heuristic player.

Game	Time spent on heuristic			Relative number of simulations		
	Playout	Tree	Combi.	Playout	Tree	Combi.
battle	44.6%	0.0%	44.3%	67.2%	99.3%	66.9%
bidding-tictactoe	15.8%	0.2%	15.5%	125.4%	103.2%	125.2%
blocker	45.3%	0.1%	42.8%	65.0%	95.9%	70.2%
breakthrough	53.4%	0.1%	53.2%	46.3%	104.4%	46.2%
checkers-small	22.9%	0.1%	22.9%	80.6%	103.0%	81.4%
chinesecheckers2	8.3%	0.1%	8.4%	91.9%	101.0%	91.1%
chinook	9.2%	0.0%	9.2%	129.6%	102.7%	131.4%
connect4	56.1%	1.6%	55.0%	91.2%	115.8%	96.7%
crisscross	5.5%	0.6%	6.0%	100.4%	99.8%	103.2%
ghostmaze	44.4%	0.7%	43.3%	70.7%	102.9%	74.9%
9BoardTicTacToe	47.4%	0.6%	46.3%	174.7%	103.6%	176.4%
pentago_2008	69.4%	0.3%	69.3%	52.2%	100.2%	52.2%
sheep_and_wolf	14.1%	0.1%	14.1%	84.4%	103.0%	84.1%
skirmish	26.3%	0.1%	26.2%	72.2%	100.5%	73.0%

5.2 Time Spent on Evaluating the Heuristic

Table 3 shows how much time during the game time was spent on evaluating the heuristic and the time needed to generate the heuristic functions for each game. These numbers do not include the time required for grounding the game description. However, most general game players use grounded game descriptions for reasoning nowadays, such that grounding needs to be done anyway.

The time required to generate the action heuristic as shown in Table 4, is relatively small for the games tested, except for Checkers with almost 15 s and Battle with 13 s. However, the time spent on evaluating the heuristic is more important. While it is almost negligible for tree heuristic, it ranges from 5 to 70% depending on the game for playout and combined heuristic. The table also shows ratio between the time needed to run 10000 game simulations by the pure and the heuristic players. Surprisingly, the heuristic player can sometimes run significantly more simulations than its non-heuristic counterpart, because the heuristic makes the simulations effectively shorter and thus taking less time. A good example of this behavior is in Nine Board Tic Tac Toe, where about 50% of the game time is spent on evaluating the heuristic, but still with about three quarters more simulations done. Moreover, the combined heuristic player won 77% of the matches against pure MCTS/UCT.

An idea, how to make the evaluation faster is to investigate pruning of the heuristic formula. It seems that some parts of it are triggered only in some relatively rare game states and do not contribute much to the overall result.

Table 4. Cost of generating the action heuristic.

Game	Cost of generation
battle	13.0 s
bidding-tictactoe	0.2 s
blocker	0.2 s
breakthrough	3.5 s
checkers-small	14.8 s
chinesecheckers2	0.1 s
chinook	2.5 s
connect4	1.6 s
crisscross	2.7 s
ghostmaze	0.2 s
9BoardTicTacToe	4.0 s
pentago_2008	1.2 s
sheep_and_wolf	3.4 s
skirmish	6.8 s

Also evaluation of state fluents that are changing wildly from state to state (like control fluent) has probably a little influence on the playing strength.

5.3 Testing with Other Games

We have been able to generate action heuristic for 113 games out of 127 available on the Tiltyard server[3]. Of those 14 games that failed we were not able to ground the game rules in 5 cases. Others failed mostly because the goal condition was particularly complex.

A game that showed to be most problematic is Othello, because the grounded version of the goal is extremely big. Other games that failed include different versions of Chess, Amazons and Hex. On the other hand, there are some games, where the grounded game description is still rather big, but the goal condition itself is relatively simple. In this case, we are able to generate the action heuristic successfully. Examples of such games are Breakthrough or Skirmish.

6 Conclusion

We have presented a general method of creating action heuristic in General Game Playing based on regression and fuzzy evaluation. We used the heuristics in three different search control schemes for MCTS and demonstrated the effectiveness by comparing it with a pure MCTS/UCT, RAVE and MAST players. The combined

---

[3] tiltyard.ggp.org.

heuristic agent outperforms these players in well-known games like Bidding Tic Tac Toe, Pentago or Nine Board Tic Tac Toe; however it shows significant loss ratio in Checkers and Connect 4.

At least partly this behavior can be explained by the fact that Tic Tac Toe-like games typically contain many persistent fluents which we use to improve the heuristics. Thus, one idea for improving the quality of the heuristics in other games is to use more feature discovery techniques, such as the ones described in [12,19] or [15]. Another idea would be to regress the goal condition by more than one step. In both cases care has to be taken to not increase the evaluation time too much.

There are still certain issues to be addressed. Notably, the playout heuristic seems to be prone to be misleading. As it has the most influence on the overall performance, this behavior should be further investigated.

Future work should also investigate pruning of the heuristic formula to only include the most relevant features in order to reduce evaluation time.

References

1. Chaslot, G.M.J.B., Winands, M.H.M., Uiterwijk, J.W.H.M., van den Herik, H.J., Bouzy, B.: Progressive strategies for Monte-Carlo tree search. In: Proceedings of the 10th Joint Conference on Information Sciences (JCIS 2007), pp. 655–661. World Scientific Publishing Co. Pte. Ltd (2007)
2. Clune, J.: Heuristic evaluation functions for general game playing. In: AAAI 2007, pp. 1134–1139. AAAI Press (2007)
3. Coulom, R.: Efficient selectivity and backup operators in monte-carlo tree search. In: van den Herik, H.J., Ciancarini, P., Donkers, H.H.L.M.J. (eds.) CG 2006. LNCS, vol. 4630, pp. 72–83. Springer, Heidelberg (2007)
4. Finnsson, H.: Generalized Monte-Carlo tree search extensions for general game playing. In: Twenty-Sixth AAAI Conference on Artificial Intelligence (2012)
5. Finnsson, H., Björnsson, Y.: Simulation-based approach to general game playing. In: AAAI 2008, pp. 259–264. AAAI Press (2008)
6. Finnsson, H., Björnsson, Y.: Learning simulation control in general game playing agents. In: AAAI 2010, pp. 954–959. AAAI Press (2010)
7. Finnsson, H., Björnsson, Y.: Cadiaplayer: search-control techniques. KI $25(1)$, 9–16 (2011)
8. Gelly, S., Silver, D.: Combining online and offline knowledge in UCT. In: Proceedings of the 24th International Conference on Machine Learning. ACM International Conference Proceeding Series, vol. 227, pp. 273–280 (2007)
9. Genesereth, M.R., Love, N., Pell, B.: General game playing: overview of the AAAI competition. AI Mag. $26(2)$, 62–72 (2005)
10. Haufe, S., Schiffel, S., Thielscher, M.: Automated verification of state sequence invariants in general game playing. Artif. Intell. 187–188, 1–30 (2012)
11. Kocsis, L., Szepesvári, C.: Bandit based monte-carlo planning. In: Fürnkranz, J., Scheffer, T., Spiliopoulou, M. (eds.) ECML 2006. LNCS (LNAI), vol. 4212, pp. 282–293. Springer, Heidelberg (2006)
12. Kuhlmann, G., Dresner, K., Stone, P.: Automatic heuristic construction in a complete general game player. In: Proceedings of the Twenty-First National Conference on Artificial Intelligence, pp. 1457–1462 (2006)

13. Lanctot, M., Winands, M.H.M., Pepels, T., Sturtevant, N.R.: Monte Carlo tree search with heuristic evaluations using implicit minimax backups. In: 2014 IEEE Conference on Computational Intelligence and Games (CIG 2014), pp. 341–348. IEEE (2014)
14. Love, N., Hinrichs, T., Haley, D., Schkufza, E., Genesereth, M.: General game playing: Game description language specification. Technical report, Stanford University. http://games.stanford.edu/
15. Michulke, D., Schiffel, S.: Admissible distance heuristics for general games. In: Filipe, J., Fred, A. (eds.) ICAART 2012. CCIS, vol. 358, pp. 188–203. Springer, Heidelberg (2013)
16. Pepels, T., Tak, M.J.W., Lanctot, M., Winands, M.H.M.: Quality-based rewards for Monte-Carlo tree search simulations. In: 21st European Conference on Artificial Intelligence (ECAI 2014), pp. 705–710. IOS Press (2014)
17. Reiter, R.: Knowledge in Action: Logical Foundations for Specifying and Implementing Dynamical Systems, pp. 61–73. Massachusetts Institute of Technology, Cambridge (2001)
18. Schiffel, S., Björnsson, Y.: Efficiency of GDL reasoners. IEEE Trans. Comput. Intell. AI Games **6**(4), 343–354 (2014)
19. Schiffel, S., Thielscher, M.: Fluxplayer: a successful general game player. In: Proceedings of the 22nd AAAI Conference on Artificial Intelligence (AAAI-07), pp. 1191–1196. AAAI Press (2007)
20. Sharma, S., Kobti, Z., Goodwin, S.D.: Knowledge generation for improving simulations in UCT for general game playing. In: Wobcke, W., Zhang, M. (eds.) AI 2008. LNCS (LNAI), vol. 5360, pp. 49–55. Springer, Heidelberg (2008)
21. Tak, M.J.W., Winands, M.H.M., Björnsson, Y.: N-grams and the last-good-reply policy applied in general game playing. IEEE Trans. Comput. Intell. AI in Games **4**(2), 73–83 (2012)
22. Waledzik, K., Mandziuk, J.: An automatically generated evaluation function in general game playing. IEEE Trans. Comput. Intell. AI Games **6**(3), 258–270 (2014)
23. Winands, M.H.M., Björnsson, Y., Saito, J.T.: Monte-Carlo tree search in lines of action. IEEE Trans. Comput. Intell. AI Games **2**(4), 239–250 (2010)

Space-Consistent Game Equivalence Detection in General Game Playing

Haifeng Zhang$^{(\boxtimes)}$, Dangyi Liu, and Wenxin Li

Peking University, Beijing, China
{pkuzhf,ldy,lwx}@pku.edu.cn

Abstract. In general game playing, agents play previously unknown games by analyzing game rules which are provided in runtime. Since taking advantage of experience from past games can efficiently enhance their intelligence, it is necessary for agents to detect equivalence between games. This paper defines game equivalence formally and concentrates on a specific scale, space-consistent game equivalence (SCGE). To detect SCGE, an approach is proposed mainly reducing the complex problem to some well-studied problems. An evaluation of the approach is performed at the end.

1 Introduction

According to human experience, exploiting equivalence between a new problem and a studied problem provides a bridge for knowledge transfer, which efficiently enhances the understanding of the new problem. Therefore, for the aim of artificial intelligence, it is important to enable computer to recognize equivalence. Particularly, as a typical application of AI, it is necessary for game-playing agents to grasp the ability of detecting equivalence between games.

The main work of this paper is to discuss classification of game equivalence, define concepts of it formally and propose an approach to detect it. Since detecting the general equivalence between games is difficult, a narrowed scale of game equivalence, *space-consistent game equivalence*, is defined firstly. Then, an approach is proposed for agents to automatically detect space-consistent game equivalence, which intends to achieve an acceptable efficiency by defining a grounded rule graph and transferring the complex problem to the well-studied problems, i.e. graph isomorphism and SAT.

This paper discusses game equivalence in the domain of General Game Playing (GGP) [2], which sets up a framework for agents to play previously unknown games by being provided game rules in runtime. This framework obliges agents to take over the responsibility of analyzing game rules from human beings. The games in GGP are turn-based, synchronized and of complete information, which are described in the Game Description Language (GDL) [5].

The work of this paper can be applied to knowledge transfer between equivalent or similar games. For example, [4] introduces a method of value function transfer for speeding up reinforcement learning, based on the technique of game equivalence detection. It can also be applied to detect symmetry of games, as [8] does.

© Springer International Publishing Switzerland 2016
T. Cazenave et al. (Eds.): CGW 2015/GIGA 2015, CCIS 614, pp. 165–177, 2016.
DOI: 10.1007/978-3-319-39402-2_12

The following section provides background on GGP and introduces definitions of game. Section 3 discusses game equivalence and its narrowness. Section 4 introduces the proposed approach to detect game equivalence, which is evaluated in Sect. 5. Section 6 concludes the work of this paper.

2   General Game Playing

In the domain of General Game Playing, games are modeled as finite state machines. In this paper, the definitions of game derive from [9].

Definition 1 (Game). *Let Σ be a countable set of ground (i.e., variable-free) symbolic expressions (terms), S a set of states, and A a set of actions. A (discrete, synchronous, deterministic) game is a structure (R, s_0, T, L, u, G), where*

- $R \subseteq \Sigma$ *finite (the roles)*;
- $s_0 \in S$ *(the initial state)*;
- $T \subseteq S$ *finite (the terminal states)*;
- $L \subseteq R \times A \times S$ *finite (the legality relation)*;
- $u : (R \to A) \times S \to S$ *finite (the update function)*;
- $G \subseteq R \times \mathbb{N} \times S$ *finite (the goal relation)*.

Here, $A \subseteq \Sigma$ and $S \subseteq 2^\Sigma$. The legality relation $(r, a, s) \subseteq L$ defines action a to be a legal action for role r in state s. The update function u takes an action for each role and (synchronously) applies the joint actions to a current state, resulting in the updated state. The goal relation $(r, n, s) \subseteq G$ defines n to be the utility for role r in state s.

In General Game Playing, rules of games are described in the GDL, which is a Prolog-like language using prefix syntax. Some keywords of the GDL are defined in Table 1. As a demonstration of the GDL, the rules of Tic-tac-toe are provided in Listing 1.1.

Table 1. GDL Keywords

(*role r*)	r is a player
(*init p*)	Proposition p holds in the initial state
(*true p*)	Proposition p holds in the current state
(*legal r a*)	Player r has legal action a in the current state
(*does r a*)	Player r does Action a
(*next p*)	Proposition p holds in the next state
terminal	The current state is terminal
(*goal r n*)	Utility of player r in current terminal state is n

```
 1 (role xplayer) (role oplayer)
 2 (init (cell 1 1 b)) (init (cell 1 2 b))...(init (cell 3 3 b))
 3 (init (control xplayer))
 4 (<= (legal ?w (mark ?x ?y)) (true (cell ?x ?y b)) (true (control ?w)))
 5 (<= (legal xplayer noop) (true (control oplayer)))
 6 (<= (legal oplayer noop) (true (control xplayer)))
 7 (<= (next (cell ?m ?n x)) (does xplayer (mark ?m ?n)) (true (cell ?m ?n b)))
 8 (<= (next (cell ?m ?n o)) (does oplayer (mark ?m ?n)) (true (cell ?m ?n b)))
 9 (<= (next (cell ?m ?n ?w)) (true (cell ?m ?n ?w)) (distinct ?w b))
10 (<= (next (cell ?m ?n b)) (does ?w (mark ?j ?k)) (true (cell ?m ?n b)) (or (distinct ?m ?j)
         (distinct ?n ?k)))
11 (<= (next (control xplayer)) (true (control oplayer)))
12 (<= (next (control oplayer)) (true (control xplayer)))
13 (<= (row ?m ?x) (true (cell ?m 1 ?x)) (true (cell ?m 2 ?x)) (true (cell ?m 3 ?x)))
14 (<= (column ?n ?x) (true (cell 1 ?n ?x)) (true (cell 2 ?n ?x)) (true (cell 3 ?n ?x)))
15 (<= (diagonal ?x) (true (cell 1 1 ?x)) (true (cell 2 2 ?x)) (true (cell 3 3 ?x)))
16 (<= (diagonal ?x) (true (cell 1 3 ?x)) (true (cell 2 2 ?x)) (true (cell 3 1 ?x)))
17 (<= (line ?x) (or (row ?m ?x) (column ?m ?x) (diagonal ?x)))
18 (<= open (true (cell ?m ?n b)))
19 (<= (goal xplayer 100) (line x))
20 (<= (goal xplayer 50) (not (line x)) (not (line o)) (not open))
21 (<= (goal xplayer 0) (line o))
22 (<= (goal oplayer 100) (line o))
23 (<= (goal oplayer 50) (not (line x)) (not (line o)) (not open))
24 (<= (goal oplayer 0) (line x))
25 (<= terminal (or (line x) (line o) (not open)))
```

Listing 1.1. Rules of Tic-tac-toe

Here, the symbol $<=$ is the implication operator. Tokens starting with a question mark are variables. The first line declares two roles of the game. Lines 2–3 define the initial state. Lines 4–6 define legal actions for roles. In order to describe an asynchronous turn-based game, an extra action noop is provided to players during their opponents' turns. Lines 7–12 define the update function. For example, Line 7 implies that (cell 1 1 x) holds in the next state if xplayer does the action (mark 1 1) and (cell 1 1 b) holds in the current state. Lines 13–18 define several auxiliary propositions describing properties of the current state. It is convenient to use these propositions in the following rules. Lines 19–24 define the goal relation of the game, while Line 25 defines the terminal states.

Except the keywords and logical words, which are printed italic, all tokens are game-specific and can be replaced by other tokens without changing the meaning of the game. Auxiliary propositions and variables are used for convenience and compactness, which can be eliminated without changing the meaning of the game.

Provided a GDL description, a game is defined as follows.

Definition 2 (Game for GDL). *Let D be a valid GDL game description, whose signature determines the set of ground terms Σ. The game for D is the game (R, s_0, T, L, u, G), where*

- $R = \{r \in \Sigma | D \models$ *(role r)*$\}$
- $s_0 = \{p \in \Sigma | D \models$ *(init p)*$\}$
- $T = \{s \in S | D \cup s^{true} \models$ *terminal*$\}$
- $L = \{(r, a, s) \in R \times A \times S | D \cup s^{true} \models$ *(legal r a)*$\}$

- $u(j : R \to A, s) = \{p \in \Sigma | D \cup j^{does} \cup s^{true} \models (next\ p)\}$
- $G = \{(r, n, s) \in R \times \mathbb{N} \times S | D \cup s^{true} \models (goal\ r\ n)\}$

Here, S is defined as 2^P where $P = \{p|\ (true\ p) \in \Sigma\}$, A as $\{a|\ (legal\ r\ a) \in \Sigma\}$, s^{true} as $\{(true\ p)\ |p \in s\}$ and $(j : R \to A)^{does}$ as $\{(does\ r\ j(r))\ |r \in R\}$.

3   Game Equivalence

Two games looking different in rules may be identical in nature. [7] points out that Tic-tac-toe is identical to Number Scrabble[1]. In fact, filling the numbers of Number Scrabble into the cells of Tic-tac-toe as Fig. 1 reveals the mapping between them.

2	9	4
7	5	3
6	1	8

Fig. 1. Mapping between Tic-tac-toe and Number Scrabble. Picking a number corresponds to marking a cell, and collecting three numbers summing up to 15 corresponds to drawing a line.

Essentially, two games are equivalent exactly if the state machines described them are identical. Corresponding to the definition of game, game equivalence is defined as follows.

Definition 3 (Game Equivalence). Game $\Gamma = (R, s_0, T, L, u, G)$ and Game $\Gamma' = (R', s_0', T', L', u', G')$ (Σ and Σ' are their grounds sets, S and S' state sets, A and A' action sets, respectively) are equivalent iff there is a bijection set $\sigma = (\sigma^R : R \leftrightarrow R', \sigma^S : S \leftrightarrow S', \sigma^A : A \leftrightarrow A')$ s.t.

- $\sigma^S(s_0) = s_0'$
- $(\forall t)\ t \in T \Leftrightarrow \sigma^S(t) \in T'$
- $(\forall r, a, s)\ (r, a, s) \in L \Leftrightarrow (\sigma^R(r), \sigma^A(a), \sigma^S(s)) \in L'$
- $(\forall j : R \to A, \forall s_{cur}, s_{next} \in S)$
  $u(j, s_{cur}) = s_{next} \Leftrightarrow u'(j', \sigma^S(s_{cur})) = \sigma^S(s_{next})$, where $j' : R' \to A'$ satisfies $j'(\sigma^R(r)) = \sigma^A(j(r))$
- $(\forall r, n, s)\ (r, n, s) \in G \Leftrightarrow (\sigma^R(r), n, \sigma^S(s)) \in G'$

The bijection set σ is called a game equivalence between Γ and Γ'.

---

[1] Number Scrabble is a game for two players taking turns to pick numbers from a pool of 1–9, whose goals are collecting three numbers summing up to 15 before the opponent achieving it.

Previous works successfully detect some kinds of game equivalence. [4] proposes a rule graph to detect game equivalence caused by rules reordering and tokens scrambling. Based on it, [8] enhances the rule graph to handle arguments reordering. However, more kinds of equivalence exist, such as:

- auxiliary propositions elimination, e.g. replacing $(<= (p_0) (p_1)) (<= (p_1) (p_2))$ by $(<= (p_0) (p_2))$;
- logical conversion, e.g. replacing $(<= (\text{consequence}) (\text{not} (\text{or} (\text{condition}_1) (\text{condition}_2))))$ by $(<= (\text{consequence}) (\text{not} (\text{condition}_1)) (\text{not} (\text{condition}_2)))$;
- arguments re-encoding, e.g. replacing $(\text{true} (\text{cell } 1..3 \ 1..3 \ x))$ by $(\text{true} (\text{cell } 1..9 \ x))$.

In general, game equivalence is caused by the uncertainty of transformation from a state machine to a GDL description. A state machine can be transformed into different but equivalent propositional nets, which can be further transformed into different but equivalent GDL descriptions. Figure 2 demonstrates a reasonable sequence of transformation steps.

According to Fig. 2, each kind of game equivalence is caused by one of the steps. As to the mentioned ones, rules reordering is caused by Step 8, tokens scrambling and arguments reordering are caused by Step 7, auxiliary propositions elimination is caused by Step 5, arguments re-encoding is caused by Step 3, and logical conversion is caused by Step 2.

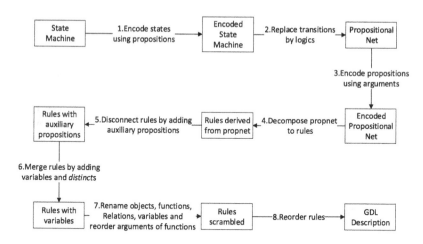

Fig. 2. Transformation steps from state machine to GDL rules. Each step can yield different targets from a single source, except for Step 4 which decomposes a propnet into certain rules.

This paper considers the steps after the encoding state machine in Fig. 2. The encoded state machines derived from a particular state machine share the same propositions. These propositions form a state space, which is also shared by the encoded state machines. To describe it, space-consistent game equivalence is defined.

Definition 4 (Space-Consistent Game Equivalence (SCGE)). *A game equivalence* $\sigma = (\sigma^R, \sigma^S, \sigma^A)$ *is a space-consistent game equivalence for two games Γ and Γ' iff there is a bijection $\sigma^P : P \leftrightarrow P'$ (where $P = \{p | p \in s, s \in S\}$, P' likewise, i.e. P and P' contains propositions forming the states) satisfying $(\forall s \in S)\, p \in s \Leftrightarrow \sigma^P(p) \in \sigma^S(s)$. Here, S is the set of states of Γ. P and P' are called state spaces. The SCGE σ can be written as $(\sigma^R, \sigma^P, \sigma^A)$, since σ^S can be determined by σ^P.*

SCGE for GDL is defined as follows, which rewrites the definition of SCGE in the context of GDL without changing the meaning.

Definition 5 (Space-Consistent Game Equivalence for GDL). *Let D and D' be valid GDL game descriptions, whose signatures determine the sets of ground terms Σ and Σ' respectively. A space-consistent game equivalence $\sigma = (\sigma^R : R \leftrightarrow R', \sigma^P : P \leftrightarrow P', \sigma^A : A \leftrightarrow A')$ for D and D' satisfies:*

- $D \models p^{init} \Leftrightarrow D' \models (\sigma^P(p))^{init}$
- $D \cup s^{true} \models terminal \Leftrightarrow D' \cup (\sigma^S(s))^{true} \models terminal$
- $D \cup s^{true} \models (legal\ r\ a) \Leftrightarrow D' \cup (\sigma^S(s))^{true} \models (legal\ \sigma^R(r)\ \sigma^A(a))$
- $D \cup s^{true} \cup \{(does\ r\ j(r))\ |r \in R\} \models (next\ p) \Leftrightarrow D' \cup (\sigma^S(s))^{true} \cup \{(does\ r\ j'(r))\ |r \in R'\} \models (next\ \sigma^P(p))$
- $D \cup s^{true} \models (goal\ r\ n) \Leftrightarrow D' \cup (\sigma^S(s))^{true} \models (goal\ \sigma^R(r)\ n)$

Here, p^{init} is defined as (init p), s^{true} as $\{(true\ p)\ |p \in s\}$ and $\sigma^S(s)$ as $\{\sigma^P(p) | p \in s\}$. $j : R \to A$ and $j' : R' \to A'$ satisfy that $(\forall r \in R)\, j'(\sigma^R(r)) = \sigma^A(j(r))$.

SCGE covers the kinds of game equivalence caused by Step 2 and after in Fig. 2. It narrows the concept of game equivalence by building a bijection between P and P' which determines the bijection between S and S', instead of building bijection between S and S' directly. For an example of space-inconsistent game equivalence which is caused by Step 1, replacing (*true* (cell 1 1 b)) in Tic-tac-toe by (not (or (*true* (cell 1 1 o)) (*true* (cell 1 1 x)))) doesn't change the game, but reduces the state space of the game.

For solving the whole problem of game equivalence detection, comparing state machines directly is a method with a very high complexity. However, comparing propositional nets covers the kinds of game equivalence caused by Step 3 and after, whose complexity is logarithmic to the corresponding state machines in most cases. Since logical conversion is a quite common situation of game equivalence, Step 2 should be also taken into consideration. This is the significance of SCGE detection.

4   Space-Consistent Game Equivalence Detection

Based on the definition of SCGE for GDL, a brute force approach to detect it is enumerating all σs mapping roles, actions and propositions of states, then checking whether all pairs of mapped terms are equivalent to each other. Specifically speaking, it consists of three phases. The first phase is generating the logical implications between keyword-propositions, which is related to the propositional net. The second phase is enumerating all possible σs mapping R to R', P to P' and A to A' so that all keyword-propositions are mapped in accordance. The third phase is verifying whether mapped keyword-propositions are equivalent to each other by comparing the logical implications generated in the first phase.

The brute force approach takes exponential time due to bijection enumeration and logical implication comparison. Therefore, Space-Consistent Game Equivalence Detection Approach (SCGEDA, or GEDA for short) is proposed. It transfers the problem to two well-studied problems, i.e. graph isomorphism and boolean satisfiability, to achieve the state-of-the-art efficiency.

The GEDA consists of three phases as the brute force approach does:

- rule grounding, which is to generate all logical implications between grounded keyword-propositions;
- graph building and mapping, which is to build a dependency graph of keyword-propositions and inspect graph isomorphisms to map keyword-propositions;
- logical equivalence verifying, which is to verify whether the mapped logical implications are equivalent.

In addition, an analysis of complexity and some efficient improvements are to be introduced.

4.1   Rule Grounding

The aim of this phase is to transfer GDL rules to equivalent rules that only contain logical implications of keyword-propositions. An example of grounded rule is displayed as follows:

$(<= (goal$ xplayer 100)
    $(or (and (true ($cell 1 1 x)) $(true ($cell 1 2 x)) $(true ($cell 1 3 x)))
       $(and (true ($cell 2 1 x)) $(true ($cell 2 2 x)) $(true ($cell 2 3 x)))
       $(and (true ($cell 3 1 x)) $(true ($cell 3 2 x)) $(true ($cell 3 3 x)))
       $(and (true ($cell 1 1 x)) $(true ($cell 2 1 x)) $(true ($cell 3 1 x)))
       $(and (true ($cell 1 2 x)) $(true ($cell 2 2 x)) $(true ($cell 3 2 x)))
       $(and (true ($cell 1 3 x)) $(true ($cell 2 3 x)) $(true ($cell 3 3 x)))
       $(and (true ($cell 1 1 x)) $(true ($cell 2 2 x)) $(true ($cell 3 3 x)))
       $(and (true ($cell 1 3 x)) $(true ($cell 2 2 x)) $(true ($cell 3 1 x)))))).
    To achieve it, several procedures are taken.

1. Calculate ranges of arguments, such as
   $(true ($cell $\{1,2,3\}$ $\{1,2,3\}$ $\{x,o\}$)).
2. Replace variables by constants according to ranges of arguments, e.g. replace
   $(<= ($diagonal ?x)
      $(true ($cell 1 1 ?x)) $(true ($cell 2 2 ?x)) $(true ($cell 3 3 ?x)))
   by

($<=$ (diagonal x)
 (*true* (cell 1 1 x)) (*true* (cell 2 2 x)) (*true* (cell 3 3 x)))
($<=$ (diagonal o)
 (*true* (cell 1 1 o)) (*true* (cell 2 2 o)) (*true* (cell 3 3 o))).

(The consistency of variables is ensured. If a *distinct*-proposition is contained in a rule, its logical value is computed and applied to the rule during this procedure.)

3. Eliminate auxiliary propositions stage by stage, e.g. replace line by row, column and diagonal before replacing row, column and diagonal by *true*.
4. Remove non-state-relative propositions and *role*-propositions from premises of rules, because their values are always true.
5. Remove the rules which use auxiliary propositions as consequences, because they are no longer of use.
6. Merge the rules which use the same propositions as consequences so that one proposition acts as the consequence in only one rule. For example, the rule with (goal xplayer 100) printed above is merged from eight partial ones with (goal xplayer 100).

After rule grounding, all rules are in the form of

($<=$ (consequence) Func(condition1, condition2, condition3...)),

where consequence and conditions are keyword-propositions. Keywords in consequence include *role, init, next, legal, goal* and *terminal*, while *true* and *does* are the keywords in conditions. Particularly, *role*- and *init*-propositions depend on no propositions as conditions, *next*-propositions depend on *true*- and *does*-propositions and the remaining consequences only depend on *true*-propositions. Func is a logical function connecting conditions by *and, or* and *not*, which is called the *reasoning function* of the consequence.

After this phase, rules of equivalent games are normalized except the reasoning functions.

4.2 Graph Building and Mapping

In this phase, the grounded rules excluding the reasoning functions are modeled as a so-called ground graph, which is mainly a dependency graph of keyword-propositions. Thus, the number of enumerated bijections between propositions is determined by the number of isomorphisms between the graphs, which is much smaller than completely enumeration.

Definition 6 (Ground Graph). *A ground graph $G = (V, E, l)$ for grounded rules GR is a directed labeled graph with the following properties:*

- *($\forall p$, p is a keyword-proposition appearing in GR with a keyword k as its predicate) $p \in V$ and $l(p) = k$;*
- *($\forall n \in \mathbb{N}, n \in [0, 100]$) $n \in V$ and $l(n) = n$;*
- *($\forall v_s, v_t \in V, r \in GR, v_s$ is a condition of r and v_t is the consequence of r) $(v_s, v_t) \in E$;*

- $(\forall p^{init}, p^{true} \in V) \, (p^{init}, p^{true}) \in E;$
- $(\forall p^{next}, p^{true} \in V) \, (p^{next}, p^{true}) \in E;$
- $(\forall a^{does,r}, r^{role} \in V) \, (a^{does,r}, r^{role}) \in E;$
- $(\forall a^{legal,r}, r^{role} \in V) \, (a^{legal,r}, r^{role}) \in E;$
- $(\forall r^{goal,n}, r^{role} \in V) \, (r^{goal,n}, r^{role}) \in E;$
- $(\forall a^{does,r}, a^{legal,r} \in V) \, (a^{does,r}, a^{legal,r}) \in E;$
- $(\forall r^{goal,n}, n \in V) \, (r^{goal,n}, n) \in E;$

Here, p^{init}, p^{true}, p^{next}, $a^{does,r}$, r^{role}, $a^{legal,r}$ and $r^{goal,n}$ express (init p), (true p), (next p), (does r a), (role r), (legal r a), (goal r n) respectively. ($\forall n \in \mathbb{N}, n \in [0, 100]$) represents all possible utilities in a valid GDL description.

Thus, a ground graph has two types of nodes, which are proposition-nodes and integer-nodes. It also has two types of edges, which are logical-dependency-edges and consistency-maintaining-edges. It only reserves logical dependencies of propositions and discards reasoning functions. Figure 3 displays a brief structure of a ground graph.

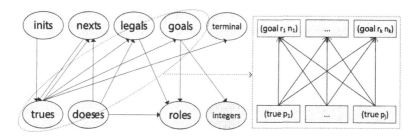

Fig. 3. Brief structure of ground graph. The solid ellipses stand for sets of nodes, while the solid rectangles stand for particular nodes.

After two ground graphs are built, they are tested for isomorphism. An isomorphism between directed labelled graphs $iso : V \leftrightarrow V'$ satisfies (1) ($\forall v \in V$) $l(v) = l'(iso(v))$; (2) ($\forall v_s, v_t \in V$) $(v_s, v_t) \in E \Leftrightarrow (iso(v_s), iso(v_t)) \in E'$.

Therefore, according to the definition of ground graph, an isomorphism between two ground graphs satisfies that (1) proposition-nodes map to proposition-nodes containing the same predicates and integer-nodes map to integer-nodes with the same value; (2) for two mapped propositions, their logically dependent propositions are also mapped; Since *init*-propositions are mapped, the initial states are equivalent; the consistencies between *next*- and *true*-propositions, *does-*, *legal-*, *goal*-propositions and *role*-propositions, *does-* and *legal*-propositions respectively are maintained; the mapped *goal*-propositions have the same utility.

After this phase, if an isomorphism is built, the two games may be equivalent. The remaining uncertainty is the reasoning functions of each proposition ,which is to be considered in the next phase.

Since a game may have symmetries [8], there may be several game equivalences between two games. In general, detecting one of them is sufficient for applications such as knowledge transfer. However, in this phase all isomorphisms of ground graphs need to be found, because any of them may cause an equivalence between games. Thus, for each isomorphism, the following phase is applied.

4.3   Logical Equivalence Verifying

In this phase, the unnormalized part of grounded rules, the reasoning functions, is handled.

By rule grounding, reasoning functions of all keyword-propositions are clear. By the last phase, mappings between keyword-propositions of two games are provided, so the reasoning functions are mapped in accordance. Moreover, propositions as conditions are also mapped. In other words, variables of reasoning functions are mapped. So, the actual problem is verifying the logical equivalence of two mapped logical functions, provided the consistent variable list. For example, there are two grounded rules $(<= $ (p1) Func(p2, p3)) and $(<= $ (q1) Func2(q2, q3)) of two games respectively, px maps qx respectively, then the problem is checking if Func(x, y) equals Func2(x, y).

For solving this problem, the naive approach that compares the truth tables of two logical functions takes exponential time. However, the problem can be transferred to the well-studied boolean satisfiability problem (SAT) to achieve a state-of-the-art efficiency. For example, testing whether logical functions f_1 and f_2 are equivalent can be transferred to testing whether $(((not\ f_1)\ and\ f_2)\ or\ (f_1\ and\ (not\ f_2)))$ is unsatisfiable. By using a SAT solver, the equivalence of two reasoning functions can be judged. So the remaining work is to verify the equivalence of all mapped reasoning functions in sequence with the SAT solver. Only if the verification is passed, the two games are equivalent and the SCGE σ can be obtained from the isomorphism of ground graphs.

4.4   Complexity

Let n be the number of reasoning functions, l the number of terms in the longest reasoning function.

For the first phase, the complexity is $O(nl)$, since the cost of grounding process is linear to the length of results.

The complexity of the second phase is at most NP-complete about n, since graph isomorphism is a special case of the NP-complete subgraph isomorphism problem [3].

The bottleneck is the third phase, which costs $O(m * n * NP - complete(l))$, where m denotes the number of maps generated by the second phase and $NP - complete(l)$ is the complexity of SAT problem [3].

The overall complexity is high. However, the approach is more efficient in practice than in theory.

Table 2. ConnectFour series games tested with their modified versions. Node No. and Edge No. express the scale of ground graph. Fun is the number of logical functions to be verified. Bij is the number of bijections generated by Phase 2 with heuristic grouping. Retry is the number of bijections verified by Phase 3 to find the first equivalence.

Game	Phase 1	Node No.	Edge No.	Phase 2	Fun	Bij	Retry	Phase 3
ConnectFour	0.107 s	217	2093	0.021 s	103	16	7	5.369 s
ConnectFourSuicide	0.100 s	217	2093	0.020 s	103	16	14	19.967 s
ConnectFourLarge	0.302 s	465	4717	0.144 s	223	64	5	8.986 s
ConnectFourLarger	1.841 s	1649	17533	9.322 s	807	1024	7	46.214 s

4.5 Improvements

There are several improvements which can be applied to the GEDA, listed by order of importance as follows.

Heuristically grouping nodes of ground graph. The number of isomorphisms of ground graphs can be huge. For example, the number of automorphisms of Tic-tac-toe's ground graph is 9!, since all 9 cells of the board are equivalent when discarding the information expressed by reasoning functions. However, the 9 cells can be grouped into 4 corner-cells, 4 border-cells and 1 center-cell by counting the numbers they are possible to form a line, which are 3, 2 and 4 respectively. This dramatically reduces the number of automorphisms to 4!4!1!. In general, analyzing the symmetry of the reasoning functions helps to group the elements of state space, so as the corresponding nodes of ground graph. Since the structure of reasoning function can be arbitrary, it is a heuristic grouping. However, it works for most occasions, because the symmetric structure is usually used by default.

Caching bad reasoning functions. Mapped reasoning functions have different possibilities to be equivalent for some reasons such as the different complexities. Caching the bad reasoning functions helps to prune early during verification.

Simplifying ground graph. Integer-nodes of ground graphs can be removed by adding a phase after graph mapping to verify the equivalence of utilities. Corresponding *true-* and *next-*proposition-nodes, *legal-* and *does-*proposition-nodes can be merged respectively. The *init-*proposition-nodes can be replaced by a single *init-*node.

Generating propositional net. Since grounded rules may need exponential space, it is more efficient to generate a propositional net and dynamically compute reasoning functions.

5 Evaluation

As introduced in Sect. 3, Tic-tac-toe is equivalent to Number Scrabble. The different part of Number Scrabble's rules is provided in Listing 1.2. The auxiliary propositions defined in Lines 3–4 represent the winning conditions. The *goal-* and *terminal-*propositions are dependent on the winning conditions. The state space

```
 1  (sum15 1 5 9) (sum15 1 6 8) (sum15 2 4 9) (sum15 2 5 8)
 2  (sum15 2 6 7) (sum15 3 4 8) (sum15 3 5 7) (sum15 4 5 6)
 3  (<= (win x) (sum15 ?a ?b ?c) (true (cell ?a x)) (true (cell ?b x)) (true (cell ?c x)))
 4  (<= (win o) (sum15 ?a ?b ?c) (true (cell ?a o)) (true (cell ?b o)) (true (cell ?c o)))
 5  (<= (goal xplayer 100) (win x))
 6  (<= (goal xplayer 50) (not (win x)) (not (win o)) (not open))
 7  (<= (goal xplayer 0) (win o))
 8  (<= (goal oplayer 100) (win o))
 9  (<= (goal oplayer 50) (not (win x)) (not (win o)) (not open))
10  (<= (goal oplayer 0) (win x))
11  (<= terminal (or (win x) (win o) (not open)))
```

Listing 1.2. Partial rules of Number Scrabble

consists of (cell [1,9] {x,o,b}) and (control {xplayer,oplayer}), which is consistent with Tic-tac-toe. Therefore, the GEDA can work on it.

Table 3. Self-mapping numbers of some games. Brute Force enumerates all permutations of the elements of state space. GEDA+ stands for the GEDA with heuristic grouping. Goal stands for the number of symmetries of a game in nature.

Game	Brute force	GEDA	GEDA+	Goal
Tic-tac-toe	27!	9!	4!4!1!	8
Blocker	48!	16!	4!4!4!4!	4
Breakthrough	128!	2	2	2
Peg jumping	66!	8	8	8
Connect four	96!	8!	2!2!2!2!	2

By the phase of rule grounding, 68 grounded rules are generated for each game.

In the phase of graph building and mapping, two ground graphs are built. Each ground graph has 59 nodes with the improvement of graph simplification. To generate isomorphisms of them, NAUTYv2.5 [6] is applied. As mentioned above, 9! isomorphisms are found between them, which can be reduced to 4!4!1!2! by the improvement of nodes grouping. The 2! is caused by the permutation of the 2 groups with 4 nodes. In fact, enumerating these isomorphisms by an agent corresponds to repeatedly trying filling the numbers in the cells by human.

For the phase of logical equivalence verifying, MiniSAT [1] is applied as a SAT solver. Since MiniSAT only accepts inputs in Conjunctive Normal Form (CNF), Tseitin transformation [10] is used to transfer the logical functions to CNF.

As a result, the equivalence of Tic-tac-toe and Number Scrabble is detected by the GEDA in 9.73 s on average over 10 experiments, running on a laptop with an Intel i5 CPU.

Since game equivalence happens rarely in nature, some manual examples are tested. Four ConnectFour series games are modified with some logical conversions. Each game is tested if it is equivalent with its modified version by the GEDA with improvements. Table 2 shows the results.

In practice, the number of enumerated bijections primarily determines the running time of a game equivalence detection approach. Table 3 displays a comparison of the enumerated self-mapping numbers of some games using different approaches, which simulate the bijection numbers between equivalent games. It reveals that the performance of the GEDA is close to the optimal for some games, while for some other games it is still unsatisfactory.

Taking into consideration that it usually takes negligible time to reject inequivalent games, the GEDA has potential to be applied in real applications.

6   Conclusion

This work makes progress toward detecting game equivalence automatically by an agent. First, it discusses the classification of game equivalence and defines the SCGE, which covers more complex game equivalences than the previous works. Second, it proposes the GEDA, which solves the problem of detecting the SCGE by using a grounded rule graph and transferring the problem to well-studied problems to achieve state-of-the-art efficiency. It works well for some small games, while there is still room for further improvement.

This work benefits knowledge transfer between equivalent games, and can be easily extended to similar games by relaxing some conditions. Based on this work, solutions which standardize state spaces of equivalent games can be proposed for space-inconsistent game equivalence detection in the future.

References

1. Een, N., Sörensson, N.: MiniSat: a SAT solver with conflict-clause minimization. SAT 5 (2005)
2. Genesereth, M., Love, N., Pell, B.: General game playing: overview of the AAAI competition. AI Mag. **26**(2), 62–72 (2005)
3. Karp, R.M.: Reducibility among combinatorial problems. In: Miller, R.E., Thatcher, J.W., Bohlinger, J.D. (eds.) Complexity of Computer Computations. Springer, New York (1972)
4. Kuhlmann, G., Stone, P.: Graph-based domain mapping for transfer learning in general games. In: Kok, J.N., Koronacki, J., Lopez de Mantaras, R., Matwin, S., Mladenič, D., Skowron, A. (eds.) ECML 2007. LNCS (LNAI), vol. 4701, pp. 188–200. Springer, Heidelberg (2007)
5. Love, N., Hinrichs, T., Haley, D., Schkufza, E., Genesereth, M.: General game playing: game description language specification (2008)
6. McKay, B.D., Piperno, A.: Practical graph isomorphism, II. J. Symb. Comput. **60**, 94–112 (2014)
7. Pell, B.: Strategy generation and evaluation for meta-game playing. Ph.D. thesis, University of Cambridge (1993)
8. Schiffel, S.: Symmetry detection in general game playing. In: AAAI (2010)
9. Schiffel, S., Thielscher, M.: A multiagent semantics for the game description language. In: Filipe, J., Fred, A., Sharp, B. (eds.) ICAART 2009. CCIS, vol. 67, pp. 44–55. Springer, Heidelberg (2010)
10. Tseitin, G.S.: On the complexity of proof in prepositional calculus. Zapiski Nauchnykh Seminarov POMI **8**, 234–259 (1968)

Author Index

Printed in the United States
By Bookmasters